MATRICES AND GRAPHS IN GEOMETRY

The main topic of this book is simplex geometry, a generalization of the geometry of the triangle and the tetrahedron. The appropriate tool for its study is matrix theory, but applications usually involve solving huge systems of linear equations or eigenvalue problems, and geometry can help in visualizing the behavior of the problem. In many cases, solving such systems may depend more on the distribution of nonzero coefficients than on their values, so graph theory is also useful. The author has discovered a method that, in many (symmetric) cases, helps to split huge systems into smaller parts.

Many readers will welcome this book, from undergraduates to specialists in mathematics, as well as nonspecialists who only use mathematics occasionally, and anyone who enjoys geometric theorems. It acquaints readers with basic matrix theory, graph theory, and elementary Euclidean geometry so that they too can appreciate the underlying connections between these various areas of mathematics and computer science.

ENCYCLOPEDIA OF MATHEMATICS AND ITS APPLICATIONS

All the titles listed below can be obtained from good booksellers or from Cambridge University Press. For a complete series listing visit

http://www.cambridge.org/uk/series/sSeries.asp?code=EOM

85 R. B. Paris and D. Kaminski *Asymptotics and Mellin–Barnes Integrals*
86 R. J. McEliece *The Theory of Information and Coding, 2nd edn*
87 B. A. Magurn *An Algebraic Introduction to K-Theory*
88 T. Mora *Solving Polynomial Equation Systems I*
89 K. Bichteler *Stochastic Integration with Jumps*
90 M. Lothaire *Algebraic Combinatorics on Words*
91 A. A. Ivanov and S. V. Shpectorov *Geometry of Sporadic Groups II*
92 P. McMullen and E. Schulte *Abstract Regular Polytopes*
93 G. Gierz *et al. Continuous Lattices and Domains*
94 S. R. Finch *Mathematical Constants*
95 Y. Jabri *The Mountain Pass Theorem*
96 G. Gasper and M. Rahman *Basic Hypergeometric Series, 2nd edn*
97 M. C. Pedicchio and W. Tholen (eds.) *Categorical Foundations*
98 M. E. H. Ismail *Classical and Quantum Orthogonal Polynomials in One Variable*
99 T. Mora *Solving Polynomial Equation Systems II*
100 E. Olivieri and M. Eulália Vares *Large Deviations and Metastability*
101 A. Kushner, V. Lychagin and V. Rubtsov *Contact Geometry and Nonlinear Differential Equations*
102 L. W. Beineke and R. J. Wilson (eds.) with P. J. Cameron *Topics in Algebraic Graph Theory*
103 O. J. Staffans *Well-Posed Linear Systems*
104 J. M. Lewis, S. Lakshmivarahan and S. K. Dhall *Dynamic Data Assimilation*
105 M. Lothaire *Applied Combinatorics on Words*
106 A. Markoe *Analytic Tomography*
107 P. A. Martin *Multiple Scattering*
108 R. A. Brualdi *Combinatorial Matrix Classes*
109 J. M. Borwein and J. D. Vanderwerff *Convex Functions*
110 M.-J. Lai and L. L. Schumaker *Spline Functions on Triangulations*
111 R. T. Curtis *Symmetric Generation of Groups*
112 H. Salzmann *et al. The Classical Fields*
113 S. Peszat and J. Zabczyk *Stochastic Partial Differential Equations with Lévy Noise*
114 J. Beck *Combinatorial Games*
115 L. Barreira and Y. Pesin *Nonuniform Hyperbolicity*
116 D. Z. Arov and H. Dym *J-Contractive Matrix Valued Functions and Related Topics*
117 R. Glowinski, J.-L. Lions and J. He *Exact and Approximate Controllability for Distributed Parameter Systems*
118 A. A. Borovkov and K. A. Borovkov *Asymptotic Analysis of Random Walks*
119 M. Deza and M. Dutour Sikirić *Geometry of Chemical Graphs*
120 T. Nishiura *Absolute Measurable Spaces*
121 M. Prest *Purity, Spectra and Localisation*
122 S. Khrushchev *Orthogonal Polynomials and Continued Fractions*
123 H. Nagamochi and T. Ibaraki *Algorithmic Aspects of Graph Connectivity*
124 F. W. King *Hilbert Transforms I*
125 F. W. King *Hilbert Transforms II*
126 O. Calin and D.-C. Chang *Sub-Riemannian Geometry*
127 M. Grabisch *et al. Aggregation Functions*
128 L. W. Beineke and R. J. Wilson (eds.) with J. L. Gross and T. W. Tucker *Topics in Topological Graph Theory*
129 J. Berstel, D. Perrin and C. Reutenauer *Codes and Automata*
130 T. G. Faticoni *Modules over Endomorphism Rings*
131 H. Morimoto *Stochastic Control and Mathematical Modeling*
132 G. Schmidt *Relational Mathematics*
133 P. Kornerup and D. W. Matula *Finite Precision Number Systems and Arithmetic*
134 Y. Crama and P. L. Hammer (eds.) *Boolean Models and Methods in Mathematics, Computer Science and Engineering*
135 V. Berthé and M. Rigo (eds.) *Combinatorics, Automata and Number Theory*
136 A. Kristály, V. D. Rădulescu and C. Varga *Variational Principles in Mathematical Physics, Geometry, and Economics*
137 J. Berstel and C. Reutenauer *Noncommutative Rational Series with Applications*
138 B. Courcelle *Graph Structure and Monadic Second-Order Logic*
139 M. Fiedler *Matrices and Graphs in Geometry*
140 N. Vakil *Real Analysis through Modern infinitesimals*

ENCYCLOPEDIA OF MATHEMATICS AND ITS APPLICATIONS

Matrices and Graphs in Geometry

MIROSLAV FIEDLER

Academy of Sciences of the Czech Republic, Prague

CAMBRIDGE
UNIVERSITY PRESS

CAMBRIDGE UNIVERSITY PRESS
Cambridge, New York, Melbourne, Madrid, Cape Town, Singapore,
São Paulo, Delhi, Dubai, Tokyo, Mexico City

Cambridge University Press
The Edinburgh Building, Cambridge CB2 8RU, UK

Published in the United States of America by Cambridge University Press, New York

www.cambridge.org
Information on this title: www.cambridge.org/9780521461931

First published 2011

Printed in the United Kingdom at the University Press, Cambridge

A catalogue record for this publication is available from the British Library

Library of Congress Cataloguing in Publication data
Fiedler, Miroslav, 1926–
Matrices and Graphs in Geometry / Miroslav Fiedler.
p. cm. – (Encyclopedia of Mathematics and its Applications; 139)
Includes bibliographical references and index.
ISBN 978-0-521-46193-1
1. Geometry. 2. Matrices. 3. Graphic methods.
I. Title. II. Series.
QA447.F45 2011
516–dc22
2010046601

ISBN 978-0-521-46193-1 Hardback

Contents

Preface

This book comprises, in addition to auxiliary material, the research on which I have worked for over 50 years. Some of the results appear here for the first time. The impetus for writing the book came from the late Victor Klee, after my talk in Minneapolis in 1991. The main subject is simplex geometry, a topic which has fascinated me since my student days, caused, in fact, by the richness of triangle and tetrahedron geometry on one side and matrix theory on the other side. A large part of the content is concerned with qualitative properties of a simplex. This can be understood as studying relations not only of equalities but also of inequalities. It seems that this direction is starting to have important consequences in practical (and important) applications, such as finite element methods.

Another feature of the book is using terminology and sometimes even more specific topics from graph theory. In fact, the interplay between Euclidean geometry, matrices, graphs, and even applications in some parts of electrical networks theory, can be considered as the basic feature of the book.

In the first chapter, the matricial methods are introduced and used for building the geometry of a simplex; the generalization of the triangle and tetrahedron to higher dimensions is also discussed. The geometric interpretations and a detailed description of basic relationships and of distinguished points in an n-simplex are given in the second chapter.

The third chapter contains a complete characterization of possible distributions of acute, right, and obtuse dihedral angles in a simplex. Also, hyperacute simplexes having no obtuse dihedral interior angle are studied. The idea of qualitative properties is extended for the position of the circumcenter of the simplex and its connection to the qualities of dihedral angles.

As can be expected, most well-known properties for the triangle allow a generalization only for special kinds of simplexes. Characterizations and deeper properties of such simplexes – right, orthocentric, cyclic, and others – are studied in the fourth chapter.

In my opinion, the methods deserve to be used not only for simplexes, but also for other geometric objects. These topics are presented in somewhat more concentrated form in Chapter 5. Let me just list them: finite sets of points, inverse simplex, simplicial cones, spherical cones, and degenerate simplexes.

The short last chapter contains some applications of the previous results. The most unusual is the remarkably close relationship of hyperacute simplexes with resistive electrical networks.

The necessary background from matrix theory, graph theory, and projective geometry is provided in the Appendix.

Miroslav Fiedler

1
A matricial approach to Euclidean geometry

1.1 Euclidean point space

We assume that the reader is familiar with the usual notion of the Euclidean vector space, i.e. a real vector space endowed by an inner product satisfying the usual conditions (cf. Appendix).

We shall be considering the *point* Euclidean n-space $\mathcal{E}_n$, which contains two kinds of objects: points and vectors. The usual operations – addition and multiplication by scalars – for vectors are here completed by analogous operations for points with the following restriction.

A linear combination of points and vectors is allowed only in two cases:

(i) the sum of the coefficients of the points is one and the result is a point;
(ii) the sum of the coefficients of the points is zero and the result is a vector.

Thus, if A and B are points, then $1.B + (-1).A$, or simply $B - A$, is a vector (which can be considered as starting at A and ending at B). The point $\frac{1}{2}A + \frac{1}{2}B$ is the midpoint of the segment AB, etc.

The points $A_0, \ldots, A_p$ are called *linearly independent* if $\alpha_0 = \ldots = \alpha_p = 0$ is the only way in which to express the zero vector as $\sum_{i=0}^{p} \alpha_i A_i$ with $\sum_{i=0}^{p} \alpha_i = 0$.

The dimension of a point Euclidean space is, by definition, the dimension of the underlying Euclidean vector space. It is equal to n if there are in the space $n+1$ linearly independent points, whereas any $n+2$ points in the space are linearly dependent.

In the usual way, we can then define linear (point) subspaces of the point Euclidean space, halfspaces, convexity, etc. A *ray*, or *halfline*, is, for some distinct points A, B, the set of all points of the form $A + \lambda(B - A)$, $\lambda \geq 0$. As usual, we define the *(Euclidean) distance* $\rho(A, B)$ between the points A, B in $\mathcal{E}_n$ as the length $\sqrt{\langle B - A, B - A\rangle}$ of the corresponding vector $B - A$. Here, as throughout the book, we denote by $\langle p, q\rangle$ the inner product of the vectors p, q in the corresponding Euclidean space.

To study geometric objects in Euclidean spaces, we shall often use positive definite and positive semidefinite matrices, or the corresponding quadratic forms. Their detailed properties will be given in the Appendix. The following basic theorem will enable us to mutually intertwine the geometric objects and matrices.

Theorem 1.1.1 *Let $p_1, \ldots, p_n$ be an ordered system of vectors in some Euclidean r-dimensional (but not $(r-1)$-dimensional) vector space. Then the Gram matrix $G(p_1, \ldots, p_n) = [g_{ik}]$, where $g_{ik} = \langle p_i, p_k \rangle$ is positive semidefinite of rank r.*

Conversely, let $A = [a_{ik}]$ be a positive semidefinite $n \times n$ matrix of rank r. Then there exists in any m-dimensional Euclidean vector space, for $m \geq r$, n vectors $p_1, \ldots, p_n$ such that

$$\langle p_i, p_k \rangle = a_{ik}, \quad \text{for all } i, k = 1, \ldots, n.$$

In addition, every linear dependence relation among the vectors p_1, $p_2, \ldots, p_n$ implies the same *linear dependence relation among the rows (and thus also columns) of the Gram matrix $G(p_1, \ldots, p_n)$, and vice-versa.*

The proof is in the Appendix, Theorems A.1.44 and A.1.45.

Another important theorem concerns so-called biorthogonal bases in the Euclidean vector space E_n. The proof will also be given in the Appendix, Theorem A.1.47.

Theorem 1.1.2 *Let $p_1, \ldots, p_n$ be an ordered system of linearly independent vectors in E_n. Then there exists a unique system of vectors $q_1, \ldots, q_n$ in E_n such that for all $i, k = 1, \ldots, n$*

$$\langle p_i, q_k \rangle = \delta_{ik}$$

(δ_{ik} is the Kronecker delta, meaning zero if $i \neq k$ and one if $i = k$). The vectors $q_1, \ldots, q_n$ are again linearly independent and the Gram matrices of the systems $p_1, \ldots, p_n$ and $q_1, \ldots, q_n$ are inverse to each other.

In other words, if $G(p)$, $G(q)$ are the Gram matrices of the vectors p_i, q_j, then the matrix

$$\begin{bmatrix} G(p) & I \\ I & G(q) \end{bmatrix}$$

has rank n.

Remark 1.1.3 The bases $p_1, \ldots, p_n$ and $q_1, \ldots, q_n$ are called *biorthogonal bases* in E_n.

We shall be using, at least in the first chapter, the usual orthonormal coordinate system in $\mathcal{E}_n$, which assigns to every point an n-tuple (usually real, but in some cases even complex) of coordinates. We also recall that the linear independence of the m points $A = (a_1, \ldots, a_n)$, $B = (b_1, \ldots, b_n)$, …, $C = (c_1, \ldots, c_n)$ is characterized by the fact that the matrix

$$\begin{bmatrix} a_1 & \cdots & a_n & 1 \\ b_1 & \cdots & b_n & 1 \\ \cdots & \cdots & \cdots & \\ c_1 & \cdots & c_n & 1 \end{bmatrix} \tag{1.1}$$

has rank m. In the case that we include the linear independence of some vector, say $u = (u_1, \ldots, u_n)$, the corresponding row in (1.1) will be $u_1, \ldots, u_n, 0$.

It then follows analogously that the *linear hull* of n linearly independent points and/or vectors, called a *hyperplane*, is determined by the relation

$$\det \begin{bmatrix} x_1 & \cdots & x_n & 1 \\ a_1 & \cdots & a_n & 1 \\ b_1 & \cdots & b_n & 1 \\ \cdots & \cdots & \cdots & \\ c_1 & \cdots & c_n & 1 \end{bmatrix} = 0.$$

This means that the point $X = (x_1, \ldots, x_n)$ is a point of this hyperplane if and only if it satisfies an equation of the form

$$\sum_{i=1}^{n} \alpha_i x_i + \alpha_0 = 0;$$

here, the n-tuple $(\alpha_1, \ldots, \alpha_n)$ cannot be a zero n-tuple because of the linear independence of the given points and/or vectors. The corresponding (thus nonzero) vector $v = [\alpha_1, \ldots, \alpha_n]^T$ (in matrix notation) is called the *normal vector* to this hyperplane. It is easily seen to be orthogonal to every vector determined by two points of the hyperplane.

Two hyperplanes are *parallel* if and only if their normal vectors are linearly dependent. They are *perpendicular* (in other words, intersect orthogonally) if the corresponding normal vectors are orthogonal. The perpendicularity is described by the formula

$$\sum_{i=1}^{n} \alpha_i \beta_i = 0 \tag{1.2}$$

if

$$\sum_{i=1}^{n} \alpha_i x_i + \alpha_0 = 0, \quad \sum_{i=1}^{n} \beta_i x_i + \beta_0 = 0$$

are equations of the two hyperplanes.

In the following chapters, it will be advantageous to use the barycentric coordinates with respect to the basic simplex.

1.2 n-simplex

An *n-simplex* in $\mathcal{E}_n$ is usually defined as the *convex hull* of $n+1$ linearly independent points, so-called *vertices*, of $\mathcal{E}_n$. (Thus a 2-simplex is a triangle, a 3-simplex a tetrahedron, etc.)

Theorem 1.2.1 *Let $A_1, \ldots, A_{n+1}$ be vertices of an n-simplex in $\mathcal{E}_n$. Then every point X in $\mathcal{E}_n$ can be expressed in the form*

$$X = \sum_{i=1}^{n+1} x_i A_i, \qquad \sum_{i=1}^{n+1} x_i = 1, \tag{1.3}$$

where the x_is are real numbers, and this expression is determined uniquely.

Also, every vector u in E_n can be expressed in the form

$$u = \sum_{i=1}^{n+1} u_i A_i, \qquad \sum u_i = 0, \tag{1.4}$$

where the u_is are real numbers, and this expression is determined uniquely.

Proof. The vectors $p_i = A_i - A_{n+1}$, $i = 1, \ldots, n$, are clearly linearly independent and thus form the basis of the corresponding vector space. Hence, if X is a point in $\mathcal{E}_n$, the vector $X - A_{n+1}$ is a linear combination of the vectors p_i: $X - A_{n+1} = \sum_{i=1}^{n} x_i p_i = \sum_{i=1}^{n} x_i(A_i - A_{n+1})$. If we denote $x_{n+1} = 1 - \sum_{i=1}^{n} x_i$, we obtain the expression in the theorem.

Suppose there is also an expression

$$X = \sum_{i=1}^{n+1} y_i A_i, \qquad \sum_{i=1}^{n+1} y_i = 1,$$

for some numbers y_i. Then for $c_i = x_i - y_i$ it would follow that

$$\sum_{i=1}^{n+1} c_i A_i = 0, \qquad \sum c_i = 0,$$

which implies $\sum_{i=1}^{n} c_i p_i = 0$. Thus $c_i = 0$, $i = 1, \ldots, n+1$, so that both expressions coincide.

If now u is a vector in E_n, then u can be written in the form

$$u = \sum_{i=1}^{n} u_i p_i = \sum_{i=1}^{n+1} u_i A_i,$$

if u_{n+1} is defined as $-\sum_{i=1}^{n} u_i$. This shows the existence of the required expression. The uniqueness follows similarly as in the first case. □

The numbers $x_1, \ldots, x_{n+1}$ in (1.3) are called *barycentric coordinates* of the point X (with respect to the n-simplex with the vertices $A_1, \ldots, A_{n+1}$). The numbers $u_1, \ldots, u_{n+1}$ in (1.4) are analogously called *barycentric coordinates* of the vector u.

It is advantageous to introduce the more general notion of homogeneous barycentric coordinates (with respect to a simplex). With their use, it is possible to study the geometric objects in the space $\overline{\mathcal{E}_n}$, i.e. the Euclidean space $\mathcal{E}_n$ completed by improper points.

Indeed, suppose that $(x_1, \ldots, x_{n+1})$ is an ordered $(n+1)$-tuple of real numbers, not all equal to zero. Distinguish two cases:

(a) $\sum_{i=1}^{n+1} x_i \neq 0$; then we assign to this $(n+1)$-tuple a *proper* point X in $\overline{\mathcal{E}_n}$ having (the usual nonhomogeneous) barycentric coordinates with respect to the given simplex

$$X = \left(\frac{x_1}{\sum x_i}, \frac{x_2}{\sum x_i}, \ldots, \frac{x_{n+1}}{\sum x_i} \right).$$

(b) $\sum_{i=1}^{n+1} x_i = 0$; then we assign to this $(n+1)$-tuple the direction of the (nonzero) vector u having (the previous) nonhomogeneous barycentric coordinates $u = (x_1, \ldots, x_{n+1})$, i.e. the *improper* point of $\overline{\mathcal{E}_n}$.

It is obvious from this definition that to every nonzero $(n+1)$-tuple $(x_1, \ldots, x_{n+1})$ a proper or improper point from $\overline{\mathcal{E}_n}$ is assigned, and to the $(n+1)$-tuples $(x_1, \ldots, x_{n+1})$ and $(\rho x_1, \ldots, \rho x_{n+1})$, for $\rho \neq 0$, the same point is assigned.

Also, conversely, to every point in $\mathcal{E}_n$ and to every direction in $\mathcal{E}_n$ a nonzero $(n+1)$-tuple of real numbers is assigned. We thus have an isomorphism between the space $\overline{\mathcal{E}_n}$ and a real projective n-dimensional space. The improper points in $\overline{\mathcal{E}_n}$ form a hyperplane with equation $\sum_{i=1}^{n+1} x_i = 0$ in the homogeneous barycentric coordinates. The points $A_1, \ldots, A_{n+1}$, i.e. the vertices of the basic simplex, have in these coordinates the form $A_1 = (1, 0, \ldots, 0), \ldots, A_{n+1} = (0, 0, \ldots, 1)$. As we shall see later, the point $(1, 1, \ldots, 1)$ is the *centroid* (barycentrum in Latin) of the simplex, which explains the name of these coordinates.

Other important objects assigned to a simplex Σ are *faces*. They are defined as linear spaces spanned by proper subsets of vertices of Σ. The word face, without specifying the dimension, is usually reserved for faces of maximum dimension, i.e. dimension $n-1$. Every such face is determined by n of the $n+1$ vertices. If A_i is the missing vertex, we denote the face by ω_i and call it the face opposite the vertex A_i. The one-dimensional faces are spanned by two vertices and are called *edges* of Σ. Sometimes, this name is assigned just to the segment between the two vertices.

It is immediately obvious that the equation of the face ω_i in barycentric coordinates is $x_i = 0$ and the smaller dimensional faces can be determined either as spans of their vertices, or as intersections of the $(n-1)$-dimensional spaces ω.

For completion, we present a lemma.

Lemma 1.2.2 *Suppose Σ is an n-simplex in $\mathcal{E}_n$ with vertices $A_1, \ldots, A_{n+1}$ and ($(n-1)$-dimensional) faces $\omega_1, \ldots, \omega_{n+1}$ (ω_i opposite A_i). The set R of those points of $\mathcal{E}_n$, not contained in any face ω_i, consists of $2^{n+1}-1$ connected open subsets, exactly one of which, called the interior of Σ, is bounded (in the sense that it does not contain any halfline). Each of these subsets is characterized by a set of signs $\varepsilon_1, \ldots, \varepsilon_{n+1}$, where each $\varepsilon_i^2 = 1$, but not all $\varepsilon_i = -1$; the proper point y, which has nonhomogeneous barycentric coordinates $y_1, \ldots, y_{n+1}$, is in this subset if and only if sgn $y_i = \varepsilon_i$, $i = 1, \ldots, n+1$. The interior of Σ consists of the points corresponding to $\varepsilon_i = 1$, $i = 1, \ldots, n+1$, thus having all barycentric coordinates positive.*

We shall not prove the whole lemma. The substantial part of the proof is in the proof of the following:

Lemma 1.2.3 *A point y is an interior point of Σ (i.e. belongs to the interior of Σ) if and only if every open halfline originating in y intersects at least one of the $(n-1)$-dimensional faces of Σ.*

Proof. Suppose first that $y = (y_1, \ldots, y_{n+1})$, $\sum y_i = 1$, is an interior point of Σ, i.e. $y_i > 0$ for $i = 1, \ldots, n+1$. If $u \neq 0$ is an arbitrary vector with nonhomogeneous barycentric coordinates $u_1, \ldots, u_{n+1}$, $\sum u_i = 0$, then the halfline $y + \lambda u$, $\lambda > 0$, necessarily intersects that face ω_k, for which k is the index such that $u_k/y_k = \min_j(u_j/y_j) < 0$ (at least one u_i is negative), namely in the point with the parameter $\lambda_0 = -y_k/u_k$.

Suppose now that $y = (y_1, \ldots, y_{n+1})$, $\sum y_i = 1$, is not an interior point of Σ. Let p be the number of positive barycentric coordinates of y, so that $0 < p < n+1$. Denote by u the vector with nonhomogeneous barycentric coordinates u_i, $i = 1, \ldots, n+1$

$$\begin{aligned} u_i &= n+1-p \quad &&\text{for } y_i > 0, \\ u_i &= -p &&\text{for } y_i \leq 0. \end{aligned}$$

We have clearly $\sum_{i=1}^{n+1} u_i = 0$, and no point of the halfline $y + \lambda u$, $\lambda > 0$, is in any face of Σ since all coordinates of all points of the halfline are different from zero. □

We now formulate the basic theorem (cf. [1]), which describes necessary and sufficient conditions for the existence of an n-simplex if the lengths of all edges are given. It generalizes the triangular inequality for the triangle.

Theorem 1.2.4 *Let $A_1, \ldots, A_{n+1}$ be vertices of an n-simplex Σ. Then the* squares $m_{ik} = |A_i - A_k|^2$ *of the lengths of its edges, $i, k = 1, \ldots, n+1$, satisfy the two conditions:*

(i) $m_{ii} = 0, \quad m_{ik} = m_{ki}$;

(ii) *for any nonzero $(n+1)$-tuple $x_1, \ldots, x_{n+1}$ of real numbers for which $\sum_{i=1}^{n+1} x_i = 0$, the inequality $\sum_{i,k=1}^{n+1} m_{ik} x_i x_k < 0$ holds.*

Conversely, if m_{ik}, $i, k = 1, \ldots, n+1$, form a system of $(n+1)^2$ real numbers satisfying the conditions (i) and (ii), then there exists in any n-dimensional Euclidean space an n-simplex with vertices $A_1, \ldots, A_{n+1}$ such that $m_{ik} = |A_i - A_k|^2$.

Proof. Suppose that $A_1, \ldots, A_{n+1}$ are vertices of an n-simplex in a Euclidean space $\mathcal{E}_n$. Then clearly (i) holds. To prove (ii), choose some orthonormal coordinate system in the underlying space. Let then $(\overset{k}{a}_1, \overset{k}{a}_2, \ldots, \overset{k}{a}_n)$ be the coordinates of A_k, $k = 1, \ldots, n+1$. Since the points $A_1, \ldots, A_{n+1}$ are linearly independent

$$\det \begin{bmatrix} \overset{1}{a}_1 & \ldots & \overset{1}{a}_n & 1 \\ \overset{2}{a}_1 & \ldots & \overset{2}{a}_n & 1 \\ \ldots & & & \\ \overset{n+1}{a}_1 & \ldots & \overset{n+1}{a}_n & 1 \end{bmatrix} \neq 0 \tag{1.5}$$

by (1.1). Suppose now that $x_1, \ldots, x_{n+1}$ is a nonzero $(n+1)$-tuple satisfying $\sum_{i=1}^{n+1} x_i = 0$. Then

$$\begin{aligned} \sum_{i,k=1}^{n+1} m_{ik} x_i x_k &= \sum_{i,k=1}^{n+1} \left(\sum_{\alpha=1}^{n} (\overset{i}{a}_\alpha - \overset{k}{a}_\alpha)^2 \right) x_i x_k \\ &= \sum_{i=1}^{n+1} \left(\sum_{\alpha=1}^{n} \overset{i}{a}_\alpha{}^2 \right) x_i \sum_{k=1}^{n+1} x_k + \sum_{i=1}^{n+1} x_i \sum_{k=1}^{n+1} \left(\sum_{\alpha=1}^{n} \overset{k}{a}_\alpha{}^2 \right) x_k \\ &\quad -2 \sum_{i,k=1}^{n+1} \sum_{\alpha=1}^{n} \overset{i}{a}_\alpha \overset{k}{a}_\alpha x_i x_k \\ &= -2 \sum_{\alpha=1}^{n} \left(\sum_{k=1}^{n+1} \overset{k}{a}_\alpha x_k \right)^2 \leq 0. \end{aligned}$$

Let us show that equality cannot be attained. In such a case, a nonzero system $x_1, \ldots, x_{n+1}$ would satisfy

$$\sum_{k=1}^{n+1} \overset{k}{a}_\alpha x_k = 0 \qquad \text{for} \quad \alpha = 1, \ldots, n,$$

and

$$\sum_{k=1}^{n+1} x_k = 0.$$

The rows of the matrix in (1.5) would thus be linearly dependent, a contradiction.

To prove the second part, assume that the numbers m_{ik} satisfy (i) and (ii). Let us show first that the numbers

$$c_{\alpha\beta} = \frac{1}{2}(m_{\alpha,n+1} + m_{\beta,n+1} - m_{\alpha\beta}), \qquad \alpha, \beta = 1, \ldots, n, \tag{1.6}$$

have the property that the quadratic form $\sum_{\alpha,\beta=1}^{n} c_{\alpha\beta}x_\alpha x_\beta$ is positive definite (and, of course, $c_{\alpha\beta} = c_{\beta\alpha}$).

Indeed, suppose that $x_1, \ldots, x_n$ is an arbitrary nonzero system of real numbers. Define $x_{n+1} = -\sum_{\alpha=1}^{n} x_\alpha$. By the assumption, $\sum_{i,k=1}^{n+1} m_{ik}x_i x_k < 0$. Now

$$\begin{aligned}
\sum_{i,k=1}^{n+1} m_{ik}x_i x_k &= \sum_{\alpha,\beta=1}^{n} m_{\alpha\beta}x_\alpha x_\beta + 2x_{n+1}\sum_{\alpha=1}^{n} m_{\alpha,n+1}x_\alpha \\
&= \sum_{\alpha,\beta=1}^{n} m_{\alpha\beta}x_\alpha x_\beta - 2\sum_{\beta=1}^{n} x_\beta \sum_{\alpha=1}^{n} m_{\alpha,n+1}x_\alpha \\
&= -2\sum_{\alpha,\beta=1}^{n} c_{\alpha\beta}x_\alpha x_\beta.
\end{aligned}$$

This implies that $\sum_{\alpha,\beta=1}^{n} c_{\alpha\beta}x_\alpha x_\beta > 0$ and the assertion about the numbers in (1.6) follows.

By Theorem 1.1.1, in an arbitrary n-dimensional Euclidean space E_n, there exist n linearly independent vectors $c_1, \ldots, c_n$ such that their inner products satisfy

$$\langle c_\alpha, c_\beta \rangle = c_{\alpha\beta}, \qquad \alpha, \beta = 1, \ldots, n.$$

Choose a point A_{n+1} in $\mathcal{E}_n$ and define points $A_1, \ldots, A_n$ by

$$A_\alpha = A_{n+1} + c_\alpha, \qquad \alpha = 1, \ldots, n.$$

Since the points $A_1, \ldots, A_{n+1}$ are linearly independent, it suffices to prove that

$$m_{ik} = |A_i - A_k|^2, \qquad i, k = 1, \ldots, n+1. \tag{1.7}$$

This holds for $k = n + 1$: $|A_\alpha - A_{n+1}|^2 = \langle c_\alpha, c_\alpha \rangle = c_{\alpha\alpha} = m_{\alpha,n+1}$ for $\alpha = 1, \ldots, n$ and, of course, also for $i = k = n + 1$. Suppose now that $i \leq n$, $k \leq n$, and $i \neq k$ (for $i = k$ (1.7) holds). Then

$$|A_i - A_k|^2 = \langle c_i - c_k, c_i - c_k\rangle = \langle c_i, c_i\rangle - 2\langle c_i, c_k\rangle + \langle c_k, c_k\rangle = m_{ik},$$

as we wanted to prove. □

Remark 1.2.5 We shall see later that $\sum_{i,k=1}^{n+1} m_{ik}x_i x_k = 0$ is the equation of the circumscribed hypersphere of the n-simplex in barycentric coordinates. The condition (ii) thus means that all the improper points are in the outer part of that hypersphere.

Remark 1.2.6 For $n = 1$, the condition in Theorem 1.2.4 means that $m_{12}x_1x_2 < 0$ whenever $x_1 + x_2 = 0$, $x \neq 0$, which is just $m_{12} > 0$. Thus positivity of all the m_{ij}s for $i \neq j$ similarly follows.

Theorem 1.2.7 *The condition* (ii) *in Theorem 1.2.4 is equivalent to the following:*

the $n \times n$ matrix $C = [c_{ik}]$, where $c_{ik} = m_{i,n+1} + m_{k,n+1} - m_{ik}$ is positive definite.

Proof. This follows by elimination of x_{n+1} from the condition (ii) in Theorem 1.2.4. □

Example 1.2.8 The Sylvester Criterion (Appendix, Theorem A.1.34) thus yields for the triangle the conditions

$$m_{13} > 0, \quad 4m_{13}m_{23} > (m_{13} + m_{23} - m_{12})^2$$

by positive definiteness of the matrix

$$\begin{bmatrix} 2m_{13} & m_{13} + m_{23} - m_{12} \\ m_{13} + m_{23} - m_{12} & 2m_{23} \end{bmatrix}.$$

From the second inequality, the usual triangle inequalities follow.

For the tetrahedron, surprisingly just three inequalities for positive definiteness of the matrix

$$\begin{bmatrix} 2m_{14} & m_{14} + m_{24} - m_{12} & m_{14} + m_{34} - m_{13} \\ m_{14} + m_{24} - m_{12} & 2m_{24} & m_{24} + m_{34} - m_{23} \\ m_{14} + m_{34} - m_{13} & m_{24} + m_{34} - m_{23} & 2m_{34} \end{bmatrix}$$

suffice.

In the sequel, we shall need some formulae for the distances and angles in barycentric coordinates.

Theorem 1.2.9 *Let $X = (x_i)$, $Y = (y_i)$, $Z = (z_i)$ be proper points in $\mathcal{E}_n$, and x_i, y_i, z_i be their homogeneous barycentric coordinates, respectively, with* *respect to the simplex Σ. Then the inner product of the vectors $Y - X$ and $Z - X$ is*

$$\langle Y - X, Z - X\rangle = -\frac{1}{2}\sum_{i,k=1}^{n+1} m_{ik}\left(\frac{x_i}{\Sigma x_j} - \frac{y_i}{\Sigma y_j}\right)\left(\frac{x_k}{\Sigma x_j} - \frac{z_k}{\Sigma z_j}\right). \tag{1.8}$$

Proof. We can assume that $\Sigma x_j = \Sigma y_j = \Sigma z_j = 1$. Then

$$\langle Y - X, Z - X\rangle = \left\langle \sum_{i=1}^{n+1}(y_i - x_i)A_i, \sum_{i=1}^{n+1}(z_i - x_i)A_i \right\rangle$$

$$= \left\langle \sum_{i=1}^{n}(y_i - x_i)(A_i - A_{n+1}), \sum_{k=1}^{n}(z_k - x_k)(A_k - A_{n+1}) \right\rangle.$$

Since

$$\langle A_i - A_{n+1}, A_k - A_{n+1}\rangle = -\frac{1}{2}(\langle A_i - A_k, A_i - A_k\rangle$$

$$-\langle A_i - A_{n+1}, A_i - A_{n+1}\rangle - \langle A_k - A_{n+1}, A_k - A_{n+1}\rangle),$$

we obtain

$$\langle Y - Z, Z - X\rangle = -\frac{1}{2}\Bigg(\sum_{i,k=1}^{n} m_{ik}(y_i - x_i)(z_k - x_k)$$

$$-\sum_{k=1}^{n}(z_k - x_k)\sum_{i=1}^{n} m_{i,n+1}(y_i - x_i) - \sum_{i=1}^{n}(y_i - x_i)\sum_{k=1}^{n} m_{k,n+1}(z_k - x_k)\Bigg)$$

$$= -\frac{1}{2}\sum_{i,k=1}^{n+1} m_{ik}(y_i - x_i)(z_k - x_k).$$

For homogeneous coordinates, this yields (1.8). □

Corollary 1.2.10 *The square of the distance between the points $X = (x_i)$ and $Y = (y_i)$ in barycentric coordinates is*

$$\rho^2(X, Y) = -\frac{1}{2}\sum_{i,k=1}^{n+1} m_{ik}\left(\frac{x_i}{\Sigma x} - \frac{y_i}{\Sigma y}\right)\left(\frac{x_k}{\Sigma x} - \frac{y_k}{\Sigma y}\right). \tag{1.9}$$

Theorem 1.2.11 *If the points $P = (p_i)$, and $Q = (q_i)$ are both improper (i.e., $\Sigma p_i = \Sigma q_i = 0$), thus corresponding to directions of lines, then these are orthogonal if and only if*

$$\sum_{i,k=1}^{n+1} m_{ik} p_i q_k = 0. \tag{1.10}$$

More generally, the cosine of the angle φ between the directions p and q satisfies

$$|\cos\varphi| = \frac{\left|\sum_{i,k=1}^{n+1} m_{ik}p_i q_k\right|}{\sqrt{\sum_{i,k=1}^{n+1} m_{ik}p_i p_k \sum_{i,k=1}^{n+1} m_{ik}q_i q_k}}. \tag{1.11}$$

Proof. Let X be an arbitrary proper point in $\mathcal{E}_n$ with barycentric coordinates x_i (so that $\sum_i x_i \neq 0$). The points Y, Z with barycentric coordinates $x_i + \lambda p_i$ (respectively, $x_i + \mu q_i$) for $\lambda \neq 0$, $\mu \neq 0$ are again proper points, and the vectors $Y - X$, $Z - X$ have the directions p and q, respectively.

The angle φ between these vectors is defined by

$$\cos\varphi = \frac{\langle Y - X, Z - X\rangle}{\sqrt{\langle Y - X, Y - X\rangle\langle Z - X, Z - X\rangle}}.$$

Substituting from (1.8), we obtain

$$\cos\varphi = \frac{(\lambda\mu)\sum_{i,k=1}^{n+1} m_{ik}p_i q_k}{\sqrt{\lambda^2\mu^2\sum_{i,k=1}^{n+1} m_{ik}p_i p_k \sum_{i,k=1}^{n+1} m_{ik}q_i q_k}},$$

which is (1.11). □

To unify these notions and use the technique of analytic *projective geometry*, we redefine $\overline{\mathcal{E}}_n$ into a projective space.

The linear independence of these "generalized" points $P = (p_1, \ldots, p_{n+1})$, $Q = (q_1, \ldots, q_{n+1}), \ldots,$ $R = (r_1, \ldots, r_{n+1})$, is reflected by the fact that the matrix

$$\begin{bmatrix} p_1 & \cdots & p_{n+1} \\ q_1 & \cdots & q_{n+1} \\ . & \cdots & . \\ r_1 & \cdots & r_{n+1} \end{bmatrix}$$

has full row-rank. This enables us to express linear dependence and to define linear subspaces. Every such linear subspace can be described either as a linear hull of points, or as the intersection of $(n-1)$-dimensional subspaces, *hyperplanes*; each hyperplane can be described as the set of all (generalized) points x, the coordinates $(x_1, \ldots, x_{n+1})$ of which satisfy a linear equality

$$\alpha_1 x_1 + \alpha_2 x_2 + \ldots + \alpha_{n+1}x_{n+1} = 0, \tag{1.12}$$

where not all coefficients $\alpha_1, \ldots, \alpha_{n+1}$ are zero. The coefficients $\alpha_1, \ldots, \alpha_{n+1}$ are (dual) coordinates of the hyperplane, and the relation (1.12) is the *incidence relation* for the point (x) and the hyperplane (α).

In accordance with (1.12)

$$\sum_{i=1}^{n+1} x_i = 0, \tag{1.13}$$

i.e. the condition that $x = (x_i)$ is improper represents the equation of a hyperplane, a so-called *improper* hyperplane.

Two (proper) hyperplanes $\sum \alpha_i x_i = 0, \sum \beta_i x_i = 0$ are different if and only if the matrix

$$\begin{bmatrix} \alpha_1 & \dots & \alpha_{n+1} \\ \beta_1 & \dots & \beta_{n+1} \end{bmatrix}$$

has rank 2. They are then parallel if and only if the rank of the matrix

$$\begin{bmatrix} \alpha_1 & \dots & \alpha_{n+1} \\ \beta_1 & \dots & \beta_{n+1} \\ 1 & \dots & 1 \end{bmatrix}$$

is again 2.

An important tool in studying the geometric properties of objects is that of using the *duality*. This can be easily studied in barycentric coordinates according to the usual duality in projective spaces (cf. Appendix, Section 7.5).

1.3 Some properties of the angles in a simplex

There is a relationship between the interior angles and the normals (i.e. lines perpendicular to the $(n-1)$-dimensional faces of an n-simplex). Let Σ be an n-simplex in $\mathcal{E}_n$ with vertices $A_1, \dots, A_{n+1}$. Denote by c_α the vectors

$$c_\alpha = A_\alpha - A_{n+1}, \quad \alpha = 1, \dots, n. \tag{1.14}$$

Theorem 1.3.1 *Let $d_1, \dots, d_n$ be an (ordered) system of vectors, which forms a biorthogonal pair of bases with the system $c_1, \dots, c_n$ from (1.14). Let d_{n+1} be the vector*

$$d_{n+1} = -\sum_{\alpha=1}^{n} d_\alpha.$$

Then the vectors

$$v_k = -d_k, \quad k = 1, \dots, n+1$$

are vectors of outer normals to the $(n-1)$-dimensional faces of Σ. The vectors v_k satisfy, and are characterized by, the relations

$$\langle A_i - A_j, v_k \rangle = -\delta_{ik} + \delta_{jk}, \quad i, j, k = 1, \dots, n+1. \tag{1.15}$$

Proof. Let $\alpha \in \{1, \dots, n\}$. Since d_i is perpendicular to all vectors c_j for $j \neq i$, d_i is orthogonal to ω_i. Let us show that $\langle d_{n+1}, c_\alpha - c_\beta \rangle = 0$ for all $\alpha, \beta \in \{1, \dots, n\}$. Indeed

$$\begin{aligned}\langle d_{n+1}, c_\alpha - c_\beta \rangle &= \Big\langle -\sum_\gamma d_\gamma, c_\alpha - c_\beta \Big\rangle \\ &= -\sum_\gamma \delta_{\gamma\alpha} + \sum_\gamma \delta_{\gamma\beta} \\ &= 0.\end{aligned}$$

Thus d_{n+1} is orthogonal to ω_{n+1}.

Let us denote by ω_k^+ (respectively, ω_k^-), $k = 1, \dots, n+1$, that halfspace determined by ω_k which contains (respectively, does not contain) the point A_k. To prove that the nonzero vector v_k is the *outer* normal of Σ, i.e. that it is directed into the halfspace ω_k^-, observe that the intersection point of ω_k with the line $A_k + \lambda d_k$ corresponds to the parameter λ_0 satisfying

$$A_k + \lambda_0 d_k = \sum_{j=1, j\neq k}^{n+1} \gamma_j A_j, \qquad \sum_{j=1, j\neq k}^{n+1} \gamma_j = 1,$$

i.e.

$$c_k + \lambda_0 d_k = \sum_{j=1, j\neq k}^{n+1} \gamma_j c_j.$$

By inner multiplication by d_k, we obtain

$$1 + \lambda_0 \langle d_k, d_k \rangle = 0.$$

Hence, $\lambda_0 < 0$, which means that this intersection point belongs to the ray corresponding to $\lambda < 0$ and v_k is the vector of the *outer* normal of Σ. Similarly

$$A_{n+1} + \lambda_0 d_{n+1} = \sum_{j=1} \gamma_j' A_j, \quad \sum_{j=1} \gamma_j' = 1,$$

determines the intersection point of ω_{n+1} with the line $A_{n+1} + \lambda d_{n+1}$. Hence

$$-\lambda_0 \sum_{\alpha=1}^n d_\alpha = \sum_{\alpha=1}^n \gamma_\alpha' c_\alpha$$

and by inner multiplication by $\sum_\alpha d_\alpha$, we obtain

$$\begin{aligned}-\lambda_0 \Big\langle \sum_\alpha d_\alpha, \sum_\alpha d_\alpha \Big\rangle &= \sum_{\alpha,\beta=1}^n \gamma_\alpha' \langle c_\alpha, d_\beta \rangle \\ &= \sum_\alpha \gamma_\alpha' \\ &= 1.\end{aligned}$$

Thus $\lambda_0 < 0$ and $-d_{n+1}$ is also the vector of an outer normal to ω_{n+1}.

The formulae (1.15) follow easily from the biorthogonality of the c_αs and d_αs and, on the other hand, are equivalent to them. □

Remark 1.3.2 We call the vectors v_k *normalized outer normals* of Σ.

It is evident geometrically, since it is essentially a planar problem, that the angle of outer normals v_i and v_k, $i \neq k$, complements the interior angle between the faces ω_i and ω_k to π, i.e. the set of all half-hyperplanes originating in the intersection $\omega_i \cup \omega_k$ and intersecting the opposite edge $A_i A_k$. We denote this angle by φ_{ik}, $i, k = 1, \ldots, n+1$.

We now use this relationship between the outer normals and interior angles for characterization of the conditions that generalize the condition that the sum of the interior angles in the triangle is π.

Theorem 1.3.3 *Let d_i be the vectors of normalized outer normals of the simplex Σ from Theorem 1.3.1. Then the interior angle φ_{ik} of the faces ω_i and ω_k $(i \neq k)$ is determined by*

$$\cos \varphi_{ik} = -\frac{\langle d_i, d_k \rangle}{\sqrt{\langle d_i, d_i \rangle} \sqrt{\langle d_k, d_k \rangle}}. \tag{1.16}$$

The matrix

$$C = \begin{bmatrix} 1 & -\cos \varphi_{12} & \ldots & -\cos \varphi_{1,n+1} \\ -\cos \varphi_{12} & 1 & \ldots & -\cos \varphi_{2,n+1} \\ \ldots & & & \\ -\cos \varphi_{1,n+1} & -\cos \varphi_{2,n+1} & \ldots & 1 \end{bmatrix} \tag{1.17}$$

then has the following properties:

(i) *its diagonal entries are all equal to one;*
(ii) *it is singular and positive semidefinite of rank n;*
(iii) *there exists a positive vector p, $p = [p_1, \ldots, p_{n+1}]^T$, such that $Cp = 0$.*

Conversely, if $C = [c_{ik}]$ is a symmetric matrix of order $n+1$ with properties (i)–(iii), *then there exists an n-simplex with interior angles φ_{ik} $(i \neq k)$ such that*

$$\cos \varphi_{ik} = -c_{ik} \qquad (i \neq k,\ i, k = 1, \ldots, n+1).$$

In addition, C is the Gram matrix of the unit vectors of outer normals of this simplex.

Proof. Equation (1.16) follows from the definition of $\varphi_{ik} = \pi - \psi_{ik}$, where ψ_{ik} is the angle spanned by the outer normals $-d_i$ and $-d_k$. To prove the properties (ii) and (iii) of C, denote by D the diagonal matrix whose diagonal entries are the numbers $\lambda_1 = \sqrt{\langle d_1, d_1\rangle}, \ldots, \lambda_{n+1} = \sqrt{\langle d_{n+1}, d_{n+1}\rangle}$. The matrix $DCD = C_1$ is clearly the Gram matrix of the system of vectors $d_1, \ldots, d_{n+1}$. Thus C_1 is positive semidefinite of rank n (since $d_1, \ldots, d_n$ are linearly independent and the row sums are equal to zero). This means that also C is positive semidefinite of rank n and, if we multiply the columns of C by the numbers $p_1 = \lambda_1, \ldots, p_{n+1} = \lambda_{n+1}$ and add together, we obtain the zero vector.

To prove the converse, suppose that $C = [c_{ik}]$ fulfills (i)–(iii). By Theorem 1.1.1, there exists in an arbitrary Euclidean n-dimensional space $\mathcal{E}_n$ a system of $n+1$ vectors $v_1, \ldots, v_{n+1}$ such that

$$\langle v_i, v_k\rangle = c_{ik},$$

and

$$\sum_{k=1}^{n+1} p_k v_k = 0. \tag{1.18}$$

Now we shall construct an n-simplex with outer normals v_i in $\mathcal{E}_n$. Choose a point U in $\mathcal{E}_n$ and define points $Q_1, \ldots, Q_{n+1}$ by $Q_i = U + v_i$, $i = 1, \ldots, n+1$. For each i, let ω_i be the hyperplane orthogonal to v_i and containing Q_i. Denote by ω_i^+ that halfspace with boundary ω_i which contains the point U. We shall show that the hyperplanes ω_i are $(n-1)$-dimensional faces of an n-simplex satisfying the conditions (i)–(iii). First, the intersection $\bigcap_i \omega_i^+$ does not contain any open halfline starting at U and not intersecting any of the hyperplanes ω_i: if the halfline $U + \lambda u$, $\lambda > 0$, did have this property for some nonzero vector u, then $\langle u, v_i\rangle \leq 0$ for $i = 1, \ldots, n+1$, thus by (1.18)

$$\sum_i p_i \langle u, v_i\rangle = 0,$$

implying $\langle u, v_i\rangle = 0$ for all i, a contradiction with the rank condition (ii).

It now follows that U is an interior point of the n-simplex and that the vectors v_i are outer normals since they satisfy (1.16). □

Remark 1.3.4 As we shall show in Chapter 2, Section 1, the numbers $p_1, \ldots, p_{n+1}$ in (iii) are proportional to the $(n-1)$-dimensional volumes of the faces (in this case convex hulls of the vertices) $\omega_1, \ldots, \omega_{n+1}$ of the simplex Σ.

1.4 Matrices assigned to a simplex

In Theorem 1.2.4, we assigned to any given n-simplex Σ an $(n+1) \times (n+1)$ matrix M, the entries of which are the *squares* of the (Euclidean) distances among the points $A_1, \ldots, A_{n+1}$

$$M = [m_{ij}], \quad m_{ij} = \rho^2(A_i, A_j), \quad i, j = 1, \ldots, n+1. \tag{1.19}$$

We call this matrix the *Menger matrix* of Σ.[1] On the other hand, denote by Q the Gram matrix of the normalized outer normals $v_1, \dots, v_{n+1}$ from (1.15)

$$Q = [\langle v_i, v_j \rangle], \quad i, j = 1, \dots, n+1. \tag{1.20}$$

We call this matrix simply the *Gramian* of Σ.

In the following theorem, we shall formulate and prove the basic relation between the matrices M and Q.

Theorem 1.4.1 *Let e be the column vector of $n+1$ ones. Then there exists a column $(n+1)$ vector $q_0 = [q_{01}, \dots, q_{0,n+1}]^T$ and a number q_{00} such that*

$$\begin{bmatrix} 0 & e^T \\ e & M \end{bmatrix} \begin{bmatrix} q_{00} & q_0^T \\ q_0 & Q \end{bmatrix} = -2I_{n+2}, \tag{1.21}$$

where I_{n+2} is the identity matrix of order $n+2$.

In other words, if we denote $Q = [q_{ik}]$, $i, k = 1, 2, \dots, n+1$, and, in addition, $m_{00} = 0$, $m_{0i} = m_{i0} = 1$, $i = 1, \dots, n+1$, then for indices $r, s, t = 0, 1, \dots, n+1$ we have

$$\sum_{s=0}^{n+1} m_{rs} q_{st} = -2\delta_{st}. \tag{1.22}$$

Proof. Partition the matrices M and Q as

$$M = \begin{bmatrix} \widetilde{M} & m \\ m^T & 0 \end{bmatrix}, \quad Q = \begin{bmatrix} \widetilde{Q} & q \\ q^T & \gamma \end{bmatrix},$$

where $\widetilde{M}, \widetilde{Q}$ are $n \times n$. Observe that by Theorem 1.3.1

$$\widetilde{Q} = [\langle v_i, v_j \rangle], \quad i, j = 1, \dots, n,$$

and

$$\widetilde{M} = [\langle c_i - c_j, c_i - c_j \rangle], \quad i, j = 1, \dots, n.$$

Since

$$\langle c_i - c_j, c_i - c_j \rangle = \langle c_i, c_i \rangle + \langle c_j, c_j \rangle - 2\langle c_i, c_j \rangle,$$

we obtain

$$[\langle c_i, c_j \rangle] = \frac{1}{2}(m\tilde{e}^T + \tilde{e}m^T - \widetilde{M}), \tag{1.23}$$

where

$$\tilde{e} = [1, \dots, 1]^T$$

with n ones.

[1] In the literature, this matrix is usually called Euclidean distance matrix.

By Theorem 1.1.2, the matrices $\widetilde{Q}$ and $[\langle c_i, c_j\rangle]$ are inverse to each other so that (1.23) implies

$$\widetilde{M}\widetilde{Q} = -2I_n + m\tilde{e}^T\widetilde{Q} + \tilde{e}m^T\widetilde{Q}. \tag{1.24}$$

Set

$$q_0 = \begin{bmatrix}\gamma \\ q\end{bmatrix},$$

where

$$q = -\widetilde{Q}m,$$

$$\gamma = -2 + \tilde{e}^T\widetilde{Q}m,$$

and

$$q_{00} = m^T\widetilde{Q}m.$$

The left-hand side of (1.21) is then (the row-sums of Q are zero)

$$\begin{bmatrix}0 & \tilde{e}^T & 1\\ \tilde{e} & \widetilde{M} & m\\ 1 & m^T & 0\end{bmatrix}\begin{bmatrix}m^T\widetilde{Q}m & -m^T\widetilde{Q} & -2+\tilde{e}^T\widetilde{Q}m\\ -\widetilde{Q}m & \widetilde{Q} & -\widetilde{Q}\tilde{e}\\ -2+\tilde{e}^T\widetilde{Q}m & -\tilde{e}^T\widetilde{Q} & \tilde{e}^T\widetilde{Q}\tilde{e}\end{bmatrix},$$

which by (1.24) is easily seen to be $-2I_{n+2}$. □

Remark 1.4.2 The relations can also be written in the following form, which will sometimes be more convenient. Denoting summation from zero to $n+1$ by indices r, s, t, summation from 1 to $n+1$ by indices i, j, k, and further $m_{0i} = m_{i0} = 1$, $i = 1, \ldots, n+1$, $m_{00} = 0$, we have (δ denoting Kronecker delta)

$$\sum_s q_{rs}m_{st} = -2\delta_{rt};$$

thus also e.g.

$$\sum_j q_{ij}m_{jk} = -q_{0j} - 2\delta_{ik}. \tag{1.25}$$

Corollary 1.4.3 *The matrix*

$$M_0 = \begin{bmatrix}0 & e^T\\ e & M\end{bmatrix} \tag{1.26}$$

is nonsingular.

The same holds for the second matrix Q_0 from (1.21) defined by

$$Q_0 = \begin{bmatrix}q_{00} & q_0^T\\ q_0 & Q\end{bmatrix}. \tag{1.27}$$

Remark 1.4.4 We call the matrix M_0 from (1.26) the *extended Menger matrix* and the matrix Q_0 from (1.27) the *extended Gramian* of Σ. It is well known (compare Appendix, (A.14)) that the determinant of M_0, which is usually called the *Cayley–Menger determinant*, is proportional to the square of the n-dimensional volume V_Σ of the simplex Σ. More precisely

$$V_\Sigma^2 = \frac{(-1)^{n+1}}{2^n (n!)^2} \det M_0. \tag{1.28}$$

It follows that

$$\text{sign}\det M_0 = (-1)^{n+1},$$

and by the formula obtained from (1.21)

$$Q_0 = (-2)M_0^{-1}, \tag{1.29}$$

along with

$$\det Q_0 < 0.$$

Remark 1.4.5 It was shown in [24] that in the formulation of Theorem 1.2.4, the part (ii) can be reformulated in terms of the extended matrix M_0 as:

(ii′) the matrix M_0 is *elliptic*, i.e. it has one eigenvalue positive and the remaining negative.

From this, it follows that in M_0 we can divide the $(i+1)$th row and column by $m_{i,n+1}$ for $i = 1, \ldots, n$ and the resulting matrix will have in its first $n+1$ rows and columns again a Menger matrix of some n-simplex.

Corollary 1.4.6 *For $I \subset N = \{1, \ldots, n+1\}$, denote by $M_0[I]$ the matrix M_0 with all rows and columns corresponding to indices in $N \setminus I$ deleted. Let $s = |I|$. Then the square of the $(s-1)$-dimensional volume $V_{\Sigma(I)}$ of the face $\Sigma(I)$ of Σ spanned by the vertices A_k, $k \in I$, is*

$$V_{\Sigma(I)}^2 = \frac{(-1)^s}{2^{s-1}((s-1)!)^2} \det M_0[I]. \tag{1.30}$$

Using the extended Gramian $\widehat{Q}$

$$V_{\Sigma(I)}^2 = -\frac{4}{((s-1)!)^2} \frac{\det Q[N \setminus I]}{\det Q_0}.$$

Here, $Q[N \setminus I]$ means the principal submatrix of Q from (1.20) *with row and column indices from $N \setminus I$.*

Proof. The first part is immediately obvious from (1.28). The second follows from (1.30) by Sylvester's identity (cf. Appendix, Theorem A.1.16) and (1.29). □

Observe that we started with an n-simplex and assigned to it the Menger matrix M and the Gram matrix Q of the normalized outer normals. In the following theorem, we shall show that we can completely reconstruct the previous situation given the matrix Q.

Theorem 1.4.7 *The following are necessary and sufficient conditions for a real $(n+1) \times (n+1)$-matrix Q to be the Gramian of an n-simplex:*

$$Q \text{ is positive semidefinite of rank } n \tag{1.31}$$

and

$$Qe = 0. \tag{1.32}$$

Proof. By (1.21), $e^T Q = 0$ so that both conditions are clearly necessary. To show that they are sufficient, observe that the given matrix Q has positive diagonal entries, say q_{ii}. If we denote by D the diagonal matrix $\operatorname{diag}\{\sqrt{q_{11}}, \sqrt{q_{22}}, \ldots, \sqrt{q_{n+1,n+1}}\}$, then the matrix $D^{-1}QD^{-1}$ satisfies the conditions (i)–(iii) in Theorem 1.3.3, if $p = De$. By this theorem, there exists an n-simplex, the Gram matrix of the unit outer normals of which is the matrix $D^{-1}QD^{-1}$. However, this simplex has indeed the Gramian Q. □

Remark 1.4.8 Let us add a consequence of the above formulae; with q_{00} defined as above

$$\det(M - \tfrac{1}{2}q_{00}J) = 0,$$

where J is the matrix of all ones.

Theorem 1.4.9 *Let a proper hyperplane $\mathcal{H}$ have the equation $\sum_i \alpha_i x_i = 0$ in barycentric coordinates. Then the orthogonal improper point (direction) U to $\mathcal{H}$ has barycentric coordinates*

$$u_i = \sum_k q_{ik}\alpha_k. \tag{1.33}$$

If, in addition, $P = (p_i)$ is a proper point, then the perpendicular line from P to $\mathcal{H}$ intersects $\mathcal{H}$ at the point $R = (r_i)$, where

$$r_i = \Big(\sum \alpha_j p_j\Big) \sum_k q_{ik}\alpha_k - \Big(\sum_{j,k} q_{jk}\alpha_j\alpha_k\Big)p_i;$$

the point symmetric to P with respect to $\mathcal{H}$ is $S = (s_i)$, where

$$s_i = 2\Big(\sum \alpha_j p_j\Big) \sum_k q_{ik}\alpha_k - \Big(\sum_{j,k} q_{jk}\alpha_j\alpha_k\Big)p_i. \tag{1.34}$$

Proof. Observe first that since $\mathcal{H}$ is proper, the numbers u_i are not all equal to zero. Now, for any $Z = (z_i)$

$$\begin{aligned}\sum_{i,k} m_{ik} u_i z_k &= \sum_{i,k,l} m_{ik} q_{il} \alpha_l z_k \\ &= -2 \sum_k \alpha_k z_k - \sum_l q_{0l} \alpha_l \sum_k z_k\end{aligned}$$

by (1.25). By (1.10), it follows that whenever Z is an improper point of $\mathcal{H}$, Z is orthogonal to U. Conversely, whenever Z is an improper point orthogonal to U, then Z belongs to $\mathcal{H}$.

It is then obvious that the point R is on the line joining P with the improper point orthogonal to $\mathcal{H}$, as well as on $\mathcal{H}$ itself. Since for non-homogeneous barycentric coordinates $r_i / \sum r_k = \frac{1}{2} p_i / \sum p_k + \frac{1}{2} s_i / \sum s_k$, R is the midpoint of PS, which completes the proof. □

Theorem 1.4.10 *The equation*

$$\alpha_0 \sum_{i,k} m_{ik} x_i x_k - 2 \sum_i \alpha_i x_i \sum_i x_i = 0 \tag{1.35}$$

is an equation of a real hypersphere in $\mathcal{E}_n$ in barycentric coordinates if the conditions

$$\alpha_0 \neq 0 \tag{1.36}$$

$$\sum_{r,s} q_{rs} \alpha_r \alpha_s > 0 \tag{1.37}$$

are fulfilled.

The center of the hypersphere has coordinates

$$c_i = \sum_r q_{ir} \alpha_r, \tag{1.38}$$

and its radius r satisfies

$$4r^2 = \frac{1}{\alpha_0^2} \sum_{r,s} q_{rs} \alpha_r \alpha_s. \tag{1.39}$$

The dual equation of the hypersphere (1.35) *is*

$$\sum_{i,k} q_{ik} \xi_i \xi_k - \frac{1}{r^2 (\sum_i c_i)^2} \Big(\sum_i c_i \xi_i \Big)^2 = 0. \tag{1.40}$$

Every real hypersphere in $\mathcal{E}_n$ has equation (1.35) *with $\alpha_0, \alpha_1, \ldots, \alpha_{n+1}$ satisfying* (1.36) *and* (1.37).

Proof. By Corollary 1.2.10, it suffices to show that for the point $C = (c_i)$ from (1.38) and the radius r from (1.39), (1.35) characterizes the condition that for the point $X = (x_i)$, $\rho(X, C) = r$. This is equivalent to the fact that for all x_is

$$\alpha_0 \sum_{i,k} m_{ik} x_i x_k - 2 \sum_i \alpha_i x_i \sum_i x_i \tag{1.41}$$
$$= -2\alpha_0 \left(-\tfrac{1}{2} \sum_{i,k} m_{ik} \left(\tfrac{x_i}{\Sigma_j x_j} - \tfrac{c_i}{\Sigma_j c_j} \right) \left(\tfrac{x_k}{\Sigma_j x_j} - \tfrac{c_k}{\Sigma_j c_j} \right) - r^2 \right) \left(\sum x_i \right)^2 .$$

Indeed, the right-hand side of (1.41) is

$$\alpha_0 \sum_{i,k} m_{ik} x_i x_k - \frac{2\alpha_0}{\sum c_j} \sum_j x_j \sum_{i,k} m_{ik} c_i x_k +$$
$$+\alpha_0 \left(\sum x_j \right)^2 \left(\frac{\sum_{i,k} m_{ik} c_i c_k}{(\sum c_j)^2} - 2r^2 \right).$$

Since by (1.21)

$$\begin{aligned} \sum_j c_j &= \alpha_0 \sum_r q_{0r} \\ &= -2\alpha_0 \end{aligned}$$

and

$$\begin{aligned} \sum_k m_{ik} c_k &= \sum_{k,r} m_{ik} q_{kr} \alpha_r \\ &= -2 \sum_r \delta_{ir} \alpha_r - m_{i0} \sum_r q_{0r} \alpha_r \\ &= -2\alpha_i - \sum_r q_{0r} \alpha_r , \end{aligned}$$

we have

$$\begin{aligned} \sum_{i,k} m_{ik} c_i c_k &= -2 \sum_{i,r} q_{ir} \alpha_i \alpha_r - 2\alpha_0 \sum_r q_{0r} \alpha_r \\ &= -2 \sum_{r,s} q_{rs} \alpha_r \alpha_s , \end{aligned}$$

as well as similarly

$$\sum_{i,k} m_{ik} c_i x_k = -2 \sum_i \alpha_i x_i - \sum_i x_i \sum_r q_{0r} \alpha_r .$$

This, together with (1.39), yields the first assertion. To find the dual equation, it suffices to invert the matrix

$$\alpha_0 M - e\alpha^T - \alpha e^T$$

of the corresponding quadratic form, α denoting the vector $[\alpha_1, \ldots, \alpha_{n+1}]^T$. It is, however, easily checked by multiplication that this is the matrix

$$-\frac{1}{2\alpha_0}\left(Q - \frac{1}{r^2(\sum_i c_i)^2} cc^T\right),$$

where $c = [c_1, \ldots, c_{n+1}]^T$.

This implies (1.40). The rest is obvious. □

Remark 1.4.11 In some cases, it is useful to generalize hyperspheres for the case that the condition (1.35) need not be satisfied. These hyperspheres, usually called *formally real*, can be considered to have purely imaginary (and nonzero) radii and play a role in generalizations of the geometry of circles. We can also define the *potency* of a proper point, say P (with barycentric coordinates p_i) with respect to the hypersphere (1.33). In elementary geometry, it is defined as $\bar{PS}^2 - r^2$, where S is the center and r the radius of the circle, in our case of the hypersphere. Using the formula (1.33), this yields the number

$$-\frac{1}{2}\frac{\sum m_{ik}p_i p_k}{(\sum p_i)^2} + \frac{\alpha_0 \sum p_i}{\sum_i \alpha_i p_i}.$$

This number is defined also in the more general case; for a usual (not formally real) hypersphere, the potency is negative for points in the *interior* of the hypersphere and positive for points outside the hypersphere. For the formally real hypersphere, the potency is positive for every proper point.

Also, we can define the angle of two hyperspheres, orthogonality, etc. Two usual (not formally real) hyperspheres with the property that the distance between their centers d and their radii r_1 and r_2 satisfies the condition $d^2 = r_1^2 + r_2^2$ are called *orthogonal*; this means that they intersect and their tangents in every common point are orthogonal. Such a property can also be defined for the formally real hyperspheres. In fact, we shall use this more general approach later when considering polarity.

Remark 1.4.12 The equation (1.35) can be obtained by elimination of a formal new indeterminate x_0 from the two equations

$$\sum_{r,s=0}^{n+1} m_{rs} x_r x_s = 0,$$

and

$$\sum_{r=0}^{n+1} \alpha_r x_r = 0.$$

Corollary 1.4.13 *The equation of the circumscribed sphere to the simplex Σ is*

$$\sum_{i,k=1}^{n+1} m_{ik} x_i x_k = 0. \tag{1.42}$$

Its center, circumcenter, *is the point (q_{0i}), and the square of its radius is $\frac{1}{4} q_{00}$, where the q_{0j}s satisfy* (1.21).

Proof. By Theorem 1.4.10 applied to $\alpha_0 = 1$, $\alpha_1 = \ldots = \alpha_{n+1} = 0$, (1.42) is the equation of a real sphere with center q_{0i} and where the square of the radius is $\frac{1}{4} q_{00}$. (The conditions (1.36) and (1.37) are satisfied.) Since $m_{ii} = 0$ for $i = 1, \ldots, n+1$, the hypersphere contains all vertices of the simplex Σ. This proves the assertion. In addition, (1.42) is – up to a nonzero factor – the equation of the only hypersphere containing all the vertices of Σ. □

Corollary 1.4.14 *Let $A = (a_i)$ be a proper point in $\widehat{\mathcal{E}}_n$, and let $\alpha \equiv \sum_{i=1}^{n+1} \alpha_i x_i = 0$ be the equation of a hyperplane. Then the distance of A from α is given by*

$$\rho(A, \alpha) = \left| \frac{\sum_{i=1}^{n+1} a_i \alpha_i}{\sqrt{\sum_{i,k=1}^{n+1} q_{ik} \alpha_i \alpha_k} \sum_{i=1}^{n+1} a_i} \right| . \tag{1.43}$$

In particular, $1/\sqrt{q_{ii}}$ is the height (i.e. the length of the altitude of Σ) corresponding to A_i.

Proof. By (1.40) in Theorem 1.4.10, the dual equation of the hypersphere with center A and radius r is

$$\sum_{i,k} q_{ik} \xi_i \xi_k - \frac{1}{r^2 (\sum_i a_i)^2} \Big(\sum_i a_i \xi_i \Big)^2 = 0.$$

If we substitute α_i for ξ_i, this yields the condition that r is the distance considered. Since then $r = \rho(A, \alpha)$ from (1.43), the proof is complete. □

2
Simplex geometry

2.1 Geometric interpretations

We start by recalling the basic philosophy of constructions in simplex geometry. In triangle geometry, we can perform constructions from the lengths of segments and magnitudes of plane angles. In solid geometry, we can proceed similarly, adding angles between planes. However, one of the simplest tasks, constructing the tetrahedron when the lengths of all its edges are given, requires using circles in space. Among the given quantities, we usually do not have areas of faces, and never the sine of a space angle. We shall be using lengths only and the usual angles. All existence questions will be transferred to the basic theorem (Theorem 1.2.4) on the existence of a simplex with given lengths of edges.

Corollary 1.4.13 allows us to completely characterize the entries of the Gramian and the extended Gramian of the n-simplex Σ.

Theorem 2.1.1 *Let $Q = [q_{ij}]$, $i, j = 1, \ldots, n+1$, be the Gramian of the n-simplex Σ, i.e. the matrix from (1.20), and $\hat{Q} = [q_{rs}]$, $r, s = 0, 1, \ldots, n+1$ the extended Gramian of Σ. The entries q_{rs} then have the following geometrical meaning:*

(i) *The number q_{ii}, $i = 1, \ldots, n+1$, is the reciprocal of the square of the length l_i of the altitude from A_i*

$$q_{ii} = \frac{1}{l_i^2}.$$

(ii) *For $i \neq j$, $i = 1, \ldots, n+1$*

$$q_{ij} = -\frac{\cos \varphi_{ij}}{l_i l_j},$$

where φ_{ik} denotes the interior angle between the $(n-1)$-dimensional faces ω_i and ω_k.

(iii) *The number* q_{00} *is equal to* $4r^2$, r *being the radius of the circumscribed hypersphere.*

(iv) *The number* q_{0i} *is the* (-2)*-multiple of the nonhomogeneous* ith *barycentric coordinate of the circumcenter.*

Proof. (i) was already discussed in Corollary 1.4.14; (ii) is a consequence of (1.16); (iii) and (iv) follow from Corollary 1.4.13 and the fact that $\sum q_{0i} = -2$ by (1.21). □

Let us illustrate these facts with the example of the triangle.

Example 2.1.2 Let ABC be a triangle with usual parameters: lengths of sides a, b, and c; angles α, β, and γ. The extended Menger matrix is then

$$M_0 = \begin{bmatrix} 0 & 1 & 1 & 1 \\ 1 & 0 & c^2 & b^2 \\ 1 & c^2 & 0 & a^2 \\ 1 & b^2 & a^2 & 0 \end{bmatrix},$$

the extended Gramian Q_0 satisfying $M_0 Q_0 = -2I$ is

$$Q_0 = \frac{1}{4S^2}\begin{bmatrix} a^2b^2c^2 & -a^2bc\cos\gamma & -ab^2c\cos\beta & -abc^2\cos\gamma \\ -a^2bc\cos\alpha & a^2 & -ab\cos\gamma & -ac\cos\beta \\ -ab^2c\cos\beta & -ab\cos\gamma & b^2 & -bc\cos\alpha \\ -abc^2\cos\gamma & -ac\cos\beta & -bc\cos\alpha & c^2 \end{bmatrix},$$

where S is the area of the triangle. We can use this fact for checking the classical theorems about the geometry of the triangle. In particular, the Heron's formula for S follows from (1.28) and the fact that

$$\det M_0 = a^4 + b^4 + c^4 - 2a^2b^2 - 2a^2c^2 - 2b^2c^2,$$

which can be decomposed as

$$-(-a+b+c)(a-b+c)(a+b-c)(a+b+c).$$

Of course, if the points A, B, and C are collinear, then

$$a^4 + b^4 + c^4 - 2a^2b^2 - 2a^2c^2 - 2b^2c^2 = 0. \tag{2.1}$$

Now, let us use Sylvester's identity (Appendix, Theorem A.1.16), the relation (1.29), and the formulae (1.28) and (1.30) to obtain further metric properties in an n-simplex.

Item (iii) in Theorem 2.1.1 allows us to express the radius r of the circumscribed hypersphere in terms of the Menger matrix as follows:

Theorem 2.1.3 *We have*

$$2r^2 = -\frac{\det M}{\det M_0}. \tag{2.2}$$

Proof. Indeed, the matrix $-\frac{1}{2}Q_0$ is the inverse matrix of M_0. By Sylvester's identity (cf. Appendix, Theorem A.1.16), $-\frac{1}{2}q_{00} = \det M/\det M_0$, which implies (2.2). □

Thus the formula (2.2) allows us to express this radius as a function of the lengths of the edges of the simplex. The same reasoning yields that

$$-\frac{1}{2}q_{ii} = \frac{\det(M_0)_i}{\det M_0},$$

where $\det(M_0)_i$ is the determinant of the submatrix of M_0 obtained by deleting the row and column with index i. Using the symbol $\mathrm{V}_n(\Sigma)$ for the n-dimensional volume of the n-simplex Σ, or, simply, $\mathrm{V}(A_1, \ldots, A_{n+1})$ if the n-simplex is determined by the vertices $A_1, \ldots, A_{n+1}$, we can see by (1.30) that

$$\frac{1}{2}q_{ii} = \frac{2^{n-1}((n-1)!)^2 \mathrm{V}_{n-1}^2(A_1, \ldots, A_{i-1}, A_{i+1}, \ldots, A_{n+1})}{2^n (n!)^2 \mathrm{V}_n^2(A_1, \ldots, A_{n+1})},$$

or

$$q_{ii} = \frac{\mathrm{V}_{n-1}^2(A_1, \ldots, A_{i-1}, A_{i+1}, \ldots, A_{n+1})}{n^2 \mathrm{V}_n^2(A_1, \ldots, A_{n+1})}.$$

Comparing this formula with (i) of Theorem 2.1.1, we obtain the expected formula

$$\mathrm{V}_n(A_1, \ldots, A_{n+1}) = \frac{1}{n} l_i \mathrm{V}_{n-1}(A_1, \ldots, A_{i-1}, A_{i+1}, \ldots, A_{n+1}). \qquad (2.3)$$

This also implies that the vector p in (iii) of Theorem 1.3.3 has the property claimed in Remark 1.3.4: that the coordinate p_i is proportional to the volume of the $(n-1)$-dimensional face opposite A_i. Indeed, the matrix C in (1.17) has by (i) and (ii) of Theorem 2.1.1 the form

$$C = DQD,$$

where D is the diagonal matrix

$$D = \mathrm{diag}(l_i).$$

Thus $Cp = 0$ implies $QDp = 0$, and since the rank of C is n, Dp is a multiple of the vector e with all ones. By (2.3), p_i is a multiple of $\mathrm{V}_{n-1}(A_1, \ldots, A_{i-1}, A_{i+1}, \ldots, A_{n+1})$ as we wanted to show; denote this volume as S_i. As before, φ_{ik} denotes the interior angle between the $(n-1)$-dimensional faces ω_i and ω_k.

Theorem 2.1.4 *The volumes S_i of the $(n-1)$-dimensional faces of an n-simplex satisfy:*

(i) $S_i = \sum_{j \neq i} S_j \cos \varphi_{ij}$ *for all* $i = 1, \ldots, n+1$;
(ii) $S_{n+1}^2 = \sum_{j=1}^n S_j^2 - 2\sum_{1 \leq j < k \leq n} S_j S_k \cos \varphi_{jk}$;
(iii) $2 \max_i S_i < \sum_i S_i$.

Proof. Since $q_{ij} = \sqrt{q_{ii}}\sqrt{q_{jj}} \cos \varphi_{ij}$, whenever $i \neq j$, the ith equation in $Qe = 0$ from (1.32) implies (i) after dividing by $\sqrt{q_{ii}}$. A simple calculation shows that

$$\begin{aligned} q_{n+1,n+1} &= -\sum_{j=1}^{n} q_{n+1,j} \\ &= \sum_{j,k=1}^{n} q_{jk}, \end{aligned}$$

which implies (ii).

The inequality (iii) follows from Theorem A.1.46 in the Appendix. □

Remark 2.1.5 Analogous results can be obtained for the reciprocals of the lengths of the altitudes using the proportionality of the l_is with the S_is.

Let us use now Sylvester's identity for the principal submatrix in the first n rows and columns of the matrix M_0. We obtain (in the same notation as in Corollary 1.4.3)

$$\frac{\det \widehat{M_0}}{\det M_0} = \frac{\sin^2 \varphi_{n,n+1}}{l_n^2 l_{n+1}^2}.$$

Therefore, by (1.30) and (1.28), if we denote the volumes by the corresponding vertices as above

$$\begin{aligned} &n\mathrm{V}_n(A_1, \ldots, A_{n+1})\mathrm{V}_{n-2}(A_1, \ldots, A_{n-1}) \\ &\quad = (n-1)\mathrm{V}_{n-1}(A_1, \ldots, A_n)\mathrm{V}_{n-1}(A_1, \ldots, A_{n-1}A_{n+1}) \sin \varphi_{n,n+1}. \end{aligned} \tag{2.4}$$

Of course, the same can be done for any two distinct indices. The formula clearly generalizes the formula for the area of the triangle; we have then to set 1 for V_0. In addition, the following holds:

Theorem 2.1.6 *An n-simplex is uniquely determined by the lengths of all edges but one and by the interior angle opposite the missing edge. If the missing edge is A_1A_2, then this simplex exists if there exist both the simplexes with the sets of vertices $A_1, A_3, \ldots, A_{n+1}$ and $A_2, \ldots, A_{n+1}$.*

Proof. Suppose that the n-simplex Σ with the extended Menger matrix M_0 and extended Gramian Q_0 as well as the n-simplex Σ' with matrices M_0' and Q_0' have both the properties so that $m_{ik} = m_{ik}'$ for all $i, k,\ i < k,\ (i, k) \neq (1, 2)$, and $\varphi_{12} = \varphi_{12}'$. Since by (A.3) in the Appendix for the adjoints

$$-\frac{1}{2}Q_0 = \frac{\mathrm{adj} M_0}{\det M_0}, \quad -\frac{1}{2}Q_0' = \frac{\mathrm{adj} M_0'}{\det M_0'}, \tag{2.5}$$

it follows that

$$q_{11} \det M_0 = q_{11}' \det M_0', q_{22} \det M_0 = q_{22}' \det M_0'.$$

Since by (i) and (ii) of Theorem 2.1.1

$$\frac{q_{12}}{\sqrt{q_{11}q_{22}}} = \frac{q'_{12}}{\sqrt{q'_{11}q'_{22}}},$$

it follows that also

$$q_{12} \det M_0 = q'_{12} \det M'_0.$$

Consequently, similarly as in (2.5)

$$\det \begin{bmatrix} 0 & 1 & 1 & \cdots & 1 \\ 1 & m_{12} & m_{23} & \cdots & m_{2,n+1} \\ 1 & m_{13} & 0 & \cdots & m_{3,n+1} \\ \vdots & \vdots & \vdots & \ddots & \vdots \\ 1 & m_{1,n+1} & m_{2,n+1} & \cdots & 0 \end{bmatrix}$$

$$= \det \begin{bmatrix} 0 & 1 & 1 & \cdots & 1 \\ 1 & m'_{12} & m_{23} & \cdots & m_{2,n+1} \\ 1 & m_{13} & 0 & \cdots & m_{3,n+1} \\ \vdots & \vdots & \vdots & \ddots & \vdots \\ 1 & m_{1,n+1} & m_{2,n+1} & \cdots & 0 \end{bmatrix},$$

which implies $m_{12} = m'_{12}$.

The second part is geometrically evident. □

Theorem 2.1.7 *Let φ_{ik}, $i, k = 1, \ldots, n+1$, denote the interior angles of the n-simplex. Then*

$$\det \begin{bmatrix} 1 & -\cos\varphi_{12} & \cdots & -\cos\varphi_{1,n+1} \\ -\cos\varphi_{12} & 1 & \cdots & -\cos\varphi_{2,n+1} \\ \cdots & & & \\ -\cos\varphi_{1,n+1} & -\cos\varphi_{2,n+1} & \cdots & 1 \end{bmatrix} = 0. \qquad (2.6)$$

If $\nu = \binom{n+1}{2}$ denotes the number of interior angles of the n-simplex, then any $\nu - 1$ of the interior angles determine the remaining interior angle uniquely.

Proof. The first part was proved in Theorem 1.3.3. To prove the second part, suppose that two n-simplexes Σ and Σ' have all interior angles the same, $\varphi_{ik} = \varphi'_{ik}$, with the exception of φ_{12} and φ'_{12}, for which $\varphi_{12} < \varphi'_{12}$. By (iii) of Theorem 1.3.3

$$\sum_i \xi_i^2 - \sum_{i,k,i\neq k} \xi_i \xi_k \cos\varphi_{ik} \geq 0,$$

with equality only for $\xi_i = \rho p_i$; similarly

$$\sum_i \xi_i'^2 - \sum_{i,k,i\neq k} \xi_i'\xi_k' \cos\varphi_{ik}' \geq 0,$$

with equality only for $\xi_i' = \rho' p_i'$.

Since $\cos\varphi_{ik} = \cos\varphi_{ik}'$ for $i+k>3$ and

$$\cos\varphi_{12} > \cos\varphi_{12}',$$

we have

$$\begin{aligned} 0 &\leq \sum_i p_i'^2 - \sum_{i,k,i\neq k} p_i'p_k' \cos\varphi_{ik} \\ &= \sum_i p_i'^2 - \sum_{i,k,i\neq k} p_i'p_k' \cos\varphi_{ik}' - 2p_1'p_2'(\cos\varphi_{12} - \cos\varphi_{12}') \\ &= -2p_1'p_2'(\cos\varphi_{12} - \cos\varphi_{12}') < 0, \end{aligned}$$

a contradiction.

Thus $\varphi_{12} \geq \varphi_{12}'$; analogously, $\varphi_{12}' \geq \varphi_{12}$, i.e. $\varphi_{12} = \varphi_{12}'$. □

Another geometrically evident theorem is the following:

Theorem 2.1.8 *An n-simplex is uniquely determined (up to congruence) by all edges of one of its $(n-1)$-dimensional faces and all n interior angles which this face spans with the other faces.*

We now make a conjecture:

Conjecture 2.1.9 An n-simplex is uniquely determined by the lengths of some (at least one) edges and all the interior angles opposite the not-determined edges.

Remark 2.1.10 It was shown in [8] that the conjecture holds in the case that all the given angles are right angles.

A corollary to (2.4) will be presented in Chapter 6.

As an important example, we shall consider the case of the tetrahedron in the three-dimensional Euclidean space.

Example 2.1.11 Let T be a tetrahedron with vertices A_1, A_2, A_3, and A_4. Denote for simplicity by a, b, and c the lengths of edges A_2A_3, A_1A_3, and A_1A_2, respectively, and by a', b', and c' the remaining lengths, a' opposite a, i.e. of the edge A_1A_4, etc. The extended Menger matrix M_0 of T is then

$$\begin{bmatrix} 0 & 1 & 1 & 1 & 1 \\ 1 & 0 & c^2 & b^2 & a'^2 \\ 1 & c^2 & 0 & a^2 & b'^2 \\ 1 & b^2 & a^2 & 0 & c'^2 \\ 1 & a'^2 & b'^2 & c'^2 & 0 \end{bmatrix}.$$

Denote by V the volume of T, by S_i, $i = 1, 2, 3, 4$, the area of the face opposite A_i, and by $\alpha, \beta, \gamma, \alpha', \beta', \gamma'$ the dihedral angles opposite a, b, c, a', b', c', respectively. It is convenient to denote the reciprocal of the altitude l_i from A_i by P_i, so that by (2.3), $P_i = V/(3\sqrt{2}S_i)$.

In this notation, the extended Gramian is

$$\begin{bmatrix} 4r^2 & s_1 & s_2 & s_3 & s_4 \\ s_1 & p_1^2 & -p_1p_2\cos\gamma & -p_1p_3\cos\beta & -p_1p_4\cos\alpha' \\ s_2 & -p_1p_2\cos\gamma & p_2^2 & -p_2p_3\cos\alpha & -p_2p_4\cos\beta' \\ s_3 & -p_1p_3\cos\beta & -p_2p_3\cos\alpha & p_3^2 & -p_3p_4\cos\gamma' \\ s_4 & -p_1p_4\cos\alpha' & -p_2p_4\cos\beta' & -p_3p_4\cos\gamma' & p_4^2 \end{bmatrix},$$

where r is the radius of the circumscribed hypersphere and the s_is are the barycentric coordinates of the circumcenter. As in (2.2)

$$2r^2 = -\frac{\det M}{\det M_0}.$$

Since $\det M_0 = 288V^2$ by (1.28), we obtain that

$$576r^2V^2 = -\det \begin{bmatrix} 0 & c^2 & b^2 & a'^2 \\ c^2 & 0 & a^2 & b'^2 \\ b^2 & a^2 & 0 & c'^2 \\ a'^2 & b'^2 & c'^2 & 0 \end{bmatrix}.$$

The right-hand side can be written (by appropriate multiplications of rows and columns) as

$$-\det \begin{bmatrix} 0 & cc' & bb' & aa' \\ cc' & 0 & aa' & bb' \\ bb' & aa' & 0 & cc' \\ aa' & bb' & cc' & 0 \end{bmatrix},$$

which can be expanded as $(aa'+bb'+cc')(-aa'+bb'+cc')(aa'-bb'+cc')(aa'+bb'-cc')$. We obtain the formula

$$576r^2V^2 = (aa'+bb'+cc')(-aa'+bb'+cc')(aa'-bb'+cc')(aa'+bb'-cc').$$

The formula (2.4) applied to T yields e.g.

$$3Va = 2S_1S_4\sin\alpha.$$

Thus

$$\frac{aa'}{\sin\alpha\sin\alpha'} = \frac{bb'}{\sin\beta\sin\beta'} = \frac{cc'}{\sin\gamma\sin\gamma'} = \frac{4S_1S_2S_3S_4}{9V^2},$$

which, in a sense, corresponds to the sine theorem in the triangle.

Let us return for a summary to Theorem 1.4.9, in particular (1.33). This shows that orthogonality in barycentric coordinates is the same as polarity with respect to the degenerate dual quadric with dual equation

$$\sum_{ik} q_{ik}\xi_i\xi_k = 0.$$

This quadric is a cone in the dual space with the "vertex" $(1, 1, \ldots, 1)_d$. It is called the *isotropic cone* and, as usual, the pole of every proper hyperplane $\sum_k \alpha_k x_k = 0$ is the improper point $(\sum_k q_{ik}\alpha_k)$. We use here, as well as in the sequel, the notation $(\gamma_i)_d$ for the dual point, namely the hyperplane $\sum_k \gamma_k x_k = 0$.

This also corresponds to the fact that the angle φ between two proper hyperplanes $(\alpha_i)_d$ and $(\beta_i)_d$ can be measured by

$$\cos\varphi = \frac{|\sum_{i,k} q_{ik}\alpha_i\beta_k|}{\sqrt{\sum_{i,k} q_{ik}\alpha_i\alpha_k \sum_{i,k} q_{ik}\beta_i\beta_k}}.$$

In addition, Theorem 1.4.10 can be interpreted as follows. Every hypersphere intersects the improper hyperplane in the isotropic point quadric. This is then the (nondegenerate) quadric in the $(n-1)$-dimensional improper hyperplane. We could summarize:

Theorem 2.1.12 *The metric geometry of an n-simplex is equivalent to the projective geometry of $n+1$ linearly independent points in a real projective n-dimensional space and a dual, only formally real, quadratic cone (isotropic cone) whose single real dual point is the vertex; this hyperplane (the improper hyperplane) does not contain any of the given points.*

Remark 2.1.13 In the case $n = 2$, the above-mentioned quadratic cone consists of two complex conjugate points (so-called *isotropic points*) on the line at infinity.

2.2 Distinguished objects of a simplex

In triangle geometry, there are very many geometric objects that can be considered as closely related to the triangle. Among them are distinguished points, such as the centroid, the circumcenter (the center of the circumscribed circle), the orthocenter (the intersection of all three altitudes), the incenter (the center of the inscribed circle), etc. Further important notions are distinguished lines, circles, and correspondences between points, lines, etc. In this section, we intend to generalize these objects and find analogous theorems for n-simplexes.

We already know that an n-simplex has a centroid (barycenter). It has barycentric coordinates $(\frac{1}{n+1}, \frac{1}{n+1}, \ldots, \frac{1}{n+1})$. In other words, its homogeneous barycentric coordinates are $(1, 1, \ldots, 1)$.

Another notion we already know is the circumcenter. We saw that its homogeneous barycentric coordinates are $(q_{01},\ldots,q_{0,n+1})$ by Theorem 2.1.1.

Let us already state here that an analogous notion to the orthocenter exists only in special kinds of simplexes, so-called orthocentric simplexes. We shall investigate them in a special section in Chapter 4.

We shall turn now to the incenter. By the formula (1.43), the distance of the point P with barycentric coordinates $(p_1,\ldots,p_{n+1})$ from the $(n-1)$-dimensional face ω_k is

$$\rho(P,\omega_k)=\left|\frac{p_k}{\sqrt{q_{kk}}\sum_i p_i}\right|. \tag{2.7}$$

It follows immediately that if $p_k=\sqrt{q_{kk}}$, for $k=1,\ldots,n+1$, the distances (of course, the q_{kk}s are all positive) from this point to all $(n-1)$-dimensional faces will be the same, and equal to $1/\sum_i\sqrt{q_{ii}}$. This last number is thus the radius of the inscribed hypersphere. We summarize:

Theorem 2.2.1 *The point $(\sqrt{q_{11}},\sqrt{q_{22}},\ldots,\sqrt{q_{n+1,n+1}})$ is the center of the inscribed hypersphere of Σ. The radius of the hypersphere is*

$$\frac{1}{\sum_k\sqrt{q_{kk}}}.$$

Remark 2.2.2 We can check that also all the points $P(\epsilon_1,\ldots,\epsilon_{n+1})$ with barycentric coordinates $(\epsilon_1\sqrt{q_{11}},\epsilon_2\sqrt{q_{22}},\ldots,\epsilon_{n+1}\sqrt{q_{n+1,n+1}})$ have the property that their distances from all the $(n-1)$-dimensional faces are the same. Here, the word "points" should be emphasized since it can happen that the sum of their coordinates is zero (an example is the regular tetrahedron in which of the seven possibilities of choices only four lead to points).

Theorem 2.2.3 *Suppose Σ is an n-simplex in $\mathcal{E}_n$. Then there exists a unique point L in $\mathcal{E}_n$ with the property that the sum of squares of the distances from the $(n-1)$-dimensional faces of Σ is minimal. The homogeneous barycentric coordinates of L are $(q_{11},q_{22},\ldots,q_{n+1,n+1})$. Thus it is always an interior point of the simplex.*

Proof. Let $P=(p_1,\ldots,p_{n+1})$ be an arbitrary proper point in the corresponding $\mathcal{E}_n$. The sum of squares of the distances of P to the $(n-1)$-dimensional faces of Σ satisfies by (2.7) that

$$\begin{aligned}\sum_{i=1}^{n+1}\rho^2(P,\omega_i)&=\frac{1}{\left(\sum p_i\right)^2}\sum_{i=1}^{n+1}\frac{p_i^2}{q_{ii}}\\&=\frac{1}{\sum_{i=1}^{n+1}q_{ii}}+\frac{1}{\sum_i q_{ii}(\sum p_i)^2}\left[\sum_i\frac{p_i^2}{q_{ii}}\sum_i q_{ii}-\left(\sum_i p_i\right)^2\right]\\&\geq\frac{1}{\sum_{i=1}^{n+1}q_{ii}}\end{aligned}$$

by the formula $\sum a_i^2 \sum b_i^2 - \left(\sum a_i b_i\right)^2 \geq 0$ for $a_i = p_i/\sqrt{q_{ii}}, b_i = \sqrt{q_{ii}}$. Here equality is attained if and only if $a_i = \rho b_i$, i.e. if and only if $p_i = \rho q_{ii}$. It follows that the minimum is attained if and only if P is the point $(q_{11}, \ldots, q_{n+1,n+1})$. □

Remark 2.2.4 The point L is called the *Lemoine point* of the simplex Σ. In the triangle, the Lemoine point is the intersection point of so-called symmedians. A *symmedian* is the line containing a vertex and is symmetric to the corresponding median with respect to the bisectrix. In the sequel, we shall generalize this property and define a so-called *isogonal correspondence* in an n-simplex. First, we call a point P in the Euclidean space of the n-simplex Σ a *nonboundary point*, *nb*-point for short, of Σ if P is not contained in any $(n-1)$-dimensional face of Σ. This, of course, happens if and only if all barycentric coordinates of P are different from zero.

Theorem 2.2.5 *Let Σ be an n-simplex in $\mathcal{E}_n$. Suppose that a point P with barycentric coordinates (p_i) is an nb-point of Σ. Choose any two distinct $(n-1)$-dimensional faces ω_i, ω_j $(i \neq j)$ of Σ, denote by $S_{ij}^{(1)}, S_{ij}^{(2)}$ the two hyperplanes of symmetry (bisectrices) of the faces ω_i and ω_j, and form a hyperplane ν_{ij} which is in the pencil generated by ω_i and ω_j symmetric to the hyperplane π_{ij} of the pencil containing the point P with respect to the bisectrices $S_{ij}^{(1)}$ and $S_{ij}^{(2)}$. Then all such hyperplanes ν_{ij} $(i \neq j)$ intersect at a point Q, which is again an nb-point of Σ. The (homogeneous) barycentric coordinates q_i of Q are related to the coordinates p_i of the point P by*

$$p_i q_i = \rho q_{ii}, \qquad i = 1, \ldots, n+1.$$

Proof. The equations of the hyperplanes ω_i, ω_j are $x_i = 0$, $x_j = 0$, respectively; the equations of the hyperplanes $S_{ij}^{(1)}, S_{ij}^{(2)}$ can be obtained as those of the loci of points which have the same distance from ω_i and ω_j. By (2.7), we obtain

$$\begin{aligned} x_i\sqrt{q_{jj}} - x_j\sqrt{q_{ii}} &= 0, \\ x_i\sqrt{q_{jj}} + x_j\sqrt{q_{ii}} &= 0. \end{aligned}$$

Finally, the hyperplane π_{ij} has equation

$$x_i p_j - x_j p_i = 0.$$

To determine the hyperplane ν_{ij}, observe that it is the fourth harmonic hyperplane to π_{ij} with respect to the two hyperplanes $S_{ij}^{(1)}$ and $S_{ij}^{(2)}$ (cf. Appendix, Theorem A.5.9). Thus, if

$$x_i p_j - x_j p_i = \alpha(x_i\sqrt{q_{jj}} - x_j\sqrt{q_{ii}}) + \beta(x_i\sqrt{q_{jj}} + x_j\sqrt{q_{ii}}),$$

then ν_{ij} has the form

$$\alpha(x_i\sqrt{q_{jj}} - x_j\sqrt{q_{ii}}) - \beta(x_i\sqrt{q_{jj}} + x_j\sqrt{q_{ii}}) = 0. \tag{2.8}$$

We obtain

$$\frac{p_j}{\sqrt{q_{jj}}} = \alpha + \beta, \frac{p_i}{\sqrt{q_{ii}}} = \alpha - \beta,$$

which implies that (2.8) has the form

$$x_i\sqrt{q_{jj}}\frac{p_i}{\sqrt{q_{ii}}} - x_j\sqrt{q_{ii}}\frac{p_j}{\sqrt{q_{jj}}} = 0,$$

or

$$\frac{x_i p_i}{q_{ii}} - \frac{x_j p_j}{q_{jj}} = 0.$$

Therefore, every hyperplane ν_{ij} contains the point $Q = (q_i)$ for $q_i = q_{ii}/p_i$, as asserted. The rest is obvious. □

This means that if we start in the previous construction with the point Q, we obtain the point P. The correspondence is thus an *involution* and the corresponding points can be called *isogonally conjugate.*

We have thus immediately:

Corollary 2.2.6 *The Lemoine point is isogonally conjugate to the centroid.*

Theorem 2.2.7 *Each of the centers of the hyperspheres in Theorem 2.2.1 and Remark 2.2.2 is isogonally conjugate to itself.*

Remark 2.2.8 In fact, we need not assume that both the isogonally conjugate points are proper. It is only necessary that they be *nb*-points.

Let us present another interpretation of the isogonal correspondence. We shall use the following well-known theorem about the foci of conics.

Theorem 2.2.9 *Let P and Q be distinct points in the plane. Then the locus of points X in the plane for which the sum of the distances, $PX + QX$, is constant is an ellipse; the locus of points X for which the modulus of the difference of the distances, $|PX - QX|$, is constant is a hyperbola. In both cases, P and Q are the foci of the corresponding conic.*

We shall use this theorem in the n-dimensional space, which means that (again for two points P and Q) we obtain a rotational quadric instead of a conic. In fact, we want to find the dual equation, i.e. an equation that characterizes the tangent hyperplanes of the quadric.

Theorem 2.2.10 *Let $P = (p_i)$, $Q = (q_i)$ be distinct points at least one of which is proper. Then every nonsingular rotational quadric with axis PQ and foci P and Q (in the sense that every intersection of this quadric with a plane containing PQ is a conic with foci P and Q) has the dual equation*

$$\sum q_{ik}\xi_i\xi_k - \rho\sum p_i\xi_i \sum q_i\xi_i = 0 \tag{2.9}$$

with some $\rho \neq 0$.

If both points P, Q are proper, the resulting quadric $\mathcal{Q}$ is an ellipsoid or hyperboloid according to whether the number $\rho \sum p_k \sum q_k$ is positive or negative. If one of the points is improper, the quadric is a paraboloid.

Conversely, every quadric with dual equation (2.9) *is rotational with foci P and Q.*

Proof. First, let both foci P and Q be proper. Then the quadric $\mathcal{Q}$ is the locus of points X for which either the sum of the distances $PX + QX$ (in the case of an ellipsoid), or the modulus of the difference of the distances $|PX - QX|$ (in the case of a hyperboloid) is constant, say c. In the first case, in the plane PQX, the tangent $t_1(X)$ in X to the intersection conic bisects the exterior angle of the vectors XP and XQ, whereas, in the second case, the tangent $t_2(X)$ bisects the angle PXQ itself. This implies easily that the product of distances $\rho(P, t_1(X))\rho(Q, t_1(X))$, (respectively, $\rho(P, t_2(X))\rho(Q, t_2(X))$) in both cases is independent of X, and is equal to $\frac{1}{4}(c^2 + PQ^2)$ in either case. The same is, however, true also for the distances from the tangent hyperplanes of $\mathcal{Q}$ in X.

Thus, if $\sum \xi_k x_k = 0$ is the equation of a tangent hyperplane to $\mathcal{Q}$, then by (2.9)

$$\frac{|\sum_k p_k \xi_k|}{\sqrt{\sum q_{ik}\xi_i\xi_k}|\sum p_k|}\,\frac{|\sum_k q_k \xi_k|}{\sqrt{\sum q_{ik}\xi_i\xi_k}|\sum q_k|}$$

is constant.

For the appropriate ρ, (2.9) follows. Since in the case of an ellipsoid (respectively, hyperboloid) the points P, Q are in the same (respectively, opposite) halfspace determined by the tangent hyperplane, the product $\sum \xi_k p_k \sum \xi_k q_k$ has the same (respectively, opposite) sign as $\sum p_k \sum q_k$. Thus for the ellipsoid the number $\rho \sum p_k \sum q_k$ is positive, and for the hyperboloid it is negative.

The converse in this case follows from the fact that the mentioned property of a tangent of a conic with foci P and Q is characteristic.

Suppose now that one of the points P, Q, say Q, is improper. To show that the quadric $\mathcal{Q}$, this time a rotational paraboloid with the focus P and the direction of its axis Q, has also its dual equation of the form (2.9), let $\sum_i \beta_i x_i = 0$ be the equation of the *directrix* $\mathcal{D}$ of $\mathcal{Q}$, the hyperplane for which $\mathcal{Q}$ is the locus of points having the same distance from $\mathcal{D}$ and P. A proper tangent hyperplane $\mathcal{H}$ of $\mathcal{Q}$ is then characterized by the fact that the point S symmetric to P with respect to $\mathcal{H}$ is contained in $\mathcal{D}$. Thus let $\sum_i \xi_i x_i = 0$ be the equation of $\mathcal{H}$. By (1.36)

$$2\Big(\sum_{i,j} p_j\xi_i\Big)\sum_{i,k} q_{ik}\xi_i\beta_k - \Big(\sum_{i,k} q_{ik}\xi_i\xi_k\Big)\sum_i p_i\beta_i = 0 \tag{2.10}$$

expresses the above characterization. Since P is not incident with $\mathcal{D}$, $\sum_i p_i\beta_i \neq 0$. Also, Q is orthogonal to $\mathcal{D}$ so that $q_i = \sum_k q_{ik}\beta_k$. This implies that (2.10) has indeed the form (up to a nonzero multiple) (2.9).

The converse in this case again follows similarly to the above case. □

We can give another characterization of isogonally conjugate points.

Theorem 2.2.11 *Let P be a proper nb-point of the simplex. Denote by $\overset{i}{R}$, $i = 1, \ldots, n+1$, the point symmetric to the point P with respect to the face ω_i. If the points $\overset{i}{R}$ are in a hyperplane, then they are in no other hyperplane and the direction of the vector perpendicular to this hyperplane is the (improper) isogonally conjugate point to P. If the points $\overset{i}{R}$ are not in a hyperplane, then they form vertices of a simplex and the circumcenter of this simplex is the isogonally conjugate point to P.*

Proof. As in (1.34), the points $\overset{i}{R}$ have in barycentric coordinates the form

$$\overset{i}{R} = (q_{ii}\,p_1 - 2q_{i1}\,p_i, q_{ii}\,p_2 - 2q_{i2}\,p_i, \ldots, q_{ii}\,p_{n+1} - 2q_{i,n+1}\,p_i),$$

where

$$P = (p_1, p_2, \ldots, p_{n+1}), \quad p_i \neq 0 \quad (i = 1, \ldots, n+1), \quad \sum p_i \neq 0.$$

Suppose first that the points $\overset{i}{R}$ are in some hyperplane $\gamma \equiv \sum \gamma_i x_i = 0$. Then, for $k = 1, 2, \ldots, n+1$

$$q_{kk} \sum_{i=1}^{n+1} \gamma_i\,p_i - 2p_k \sum_{i=1}^{n+1} q_{ik}\gamma_i = 0,$$

or

$$\sum_{i=1}^{n+1} q_{ik}\gamma_i = \frac{q_{kk}}{2p_k} \sum_{i=1}^{n+1} \gamma_i\,p_i. \tag{2.11}$$

The hyperplane γ is proper, i.e. the left-hand sides of the equations (2.11) are not all equal to zero. Thus $\sum \gamma_i p_i \neq 0$, and by Theorem 2.2.10 the isogonally conjugate point Q to P is the improper point of the orthogonal direction to the hyperplane γ. This also implies the uniqueness of the hyperplane γ containing all the points $\overset{i}{R}$.

Suppose now that the points $\overset{i}{R}$ are contained in no hyperplane. Then the point Q isogonally conjugate to P is proper. Indeed, otherwise the rows of the matrix $[q_{ii}\,p_j - 2q_{ij}\,p_i]$ of the coordinates of the points $\overset{i}{R}$ would be linearly dependent (after multiplication of the ith row by $\dfrac{1}{p_i}$ and summation), so that the columns would be linearly dependent, too. This is a contradiction to the fact that the points $\overset{i}{R}$ are not contained in a hyperplane. A direct computation using (2.7) and (2.10) shows that the distances of Q to all the points $\overset{i}{R}$ are equal, which completes the proof. □

We shall generalize now the isogonal correspondence.

Theorem 2.2.12 *Suppose that $A = (a_k)$ and $B = (b_k)$ are (not necessarily distinct) nb-points of Σ. For every pair of $(n-1)$-dimensional faces ω_i, ω_j $(i \neq j)$, construct a pencil of (decomposable) quadrics*

$$\lambda x_i x_j + \mu(a_i x_j - a_j x_i)(b_i x_j - b_j x_i) = 0.$$

This means that one quadric of the pencil is the pair of the faces ω_i, ω_j, the other the pair of hyperplanes (containing the intersection $\omega_i \cap \omega_j$) – one containing the point A, the other the point B. If $P = (p_i)$ is again an nb-point of Σ, then the following holds.

The quadric of the mentioned pencil, which contains the point P, decomposes into the product of the hyperplane containing the point P (and the intersection $\omega_i \cap \omega_j$) and another hyperplane H_{ij} (again containing the intersection $\omega_i \cap \omega_j$). All these hyperplanes H_{ij} have a point Q in common (for all $i, j, i \neq j$). The point Q is again an nb-point and its barycentric coordinates (q_i) have the property that

$$\frac{p_i\, q_i}{a_i\, b_i}$$

is constant.

We can thus write (using analogously to the Hadamard product of matrices or vectors the elementwise multiplication $\circ$)

$$P \circ Q = A \circ B.$$

Proof. The quadric of the mentioned pencil which contains the point P has the equation

$$\det \begin{bmatrix} (a_j x_i - a_i x_j)(b_j x_i - b_i x_j), & x_i x_j \\ (a_j p_i - a_i p_j)(b_j p_i - b_i p_j), & p_i p_j \end{bmatrix} = 0.$$

From this

$$\det \begin{bmatrix} a_j b_j x_i^2 + a_i b_i x_j^2, & x_i x_j \\ a_j b_j p_i^2 + a_i b_i p_j^2, & p_i p_j \end{bmatrix} = 0,$$

or

$$(a_j b_j x_i p_i - a_i b_i x_j p_j)(p_j x_i - p_i x_j) = 0.$$

Thus H_{ij} has the equation

$$\frac{a_j\, b_j}{p_j} x_i - \frac{a_i\, b_i}{p_i} x_j = 0$$

and all these hyperplanes have the point $Q = (q_i)$ in common, where $q_k = a_k\, b_k/p_k$, $k = 1, \ldots, n+1$. □

Corollary 2.2.13 *If the points A and B are isogonally conjugate, then so are the points P and Q.*

We can say that four *nb*-points $C = (c_i)$, $D = (d_i)$, $E = (e_i)$, $F = (f_i)$ of a simplex Σ form a *quasiparallelogram* with respect to Σ, if for some ρ

$$\frac{c_k\, e_k}{d_k\, f_k} = \rho, \qquad k = 1, \ldots, n+1.$$

The points C, E (respectively, D, F) are opposite vertices; the points C, D, etc. neighboring vertices of the quasiparallelogram.

To explain the notion, let us define a mapping of the interior of the n-simplex Σ into the n-dimensional Euclidean space $\tilde{\mathcal{E}_n}$, which is the intersection of $\mathcal{E}_{n+1}$ (with points with orthonormal coordinates $X_1, \ldots, X_{n+1}$) with the hyperplane $\sum\limits_{i=1}^{n+1} X_i = 0$ as follows: we normalize the barycentric coordinates of an arbitrary point $U = (u_1, \ldots, u_{n+1})$ of the interior of Σ (i.e., $u_i > 0$ for $i = 1, \ldots, n+1$) in such a way that $\prod\limits_{i=1}^{n+1} u_i = 1$. Then we assign to the point U the point $\tilde{U} \in \tilde{\mathcal{E}_n}$ with coordinates $U_i = \log u_i$. (Clearly, $\sum U_i = \sum \log u_i = \log \prod u_i = 0$.) In particular, to the centroid of Σ, the origin in $\tilde{\mathcal{E}_n}$ will be assigned. It is immediate that the images of the vertices of a quasiparallelogram in $\mathcal{E}_n$ will form a parallelogram (possibly degenerate) in $\tilde{\mathcal{E}_n}$, and vice-versa.

Theorem 2.2.14 *Suppose that the nb-points A, B, C, and D form a quasiparallelogram, all with respect to an n-simplex Σ. If we project these points from a vertex of Σ on the opposite face, then the resulting projections will again form a quasiparallelogram with respect to the $(n-1)$-simplex forming that face.*

Proof. This follows immediately from the fact that the projection of the point $(u_1, u_2, \ldots, u_{n+1})$ from, say, the vertex A_{n+1} has (homogeneous) barycentric coordinates $(u_1, u_2, \ldots, u_n, 0)$ in the original coordinate system, and thus $(u_1, u_2, \ldots, u_n,)$ in the coordinate system of the face. □

Remark 2.2.15 This projection can be repeated so that, in fact, the previous theorem is valid even for more general projections from one face onto the opposite face.

An important case of a quasiparallelogram occurs when the second and fourth of the vertices of the quasiparallelogram coincide with the centroid of the simplex. The remaining two vertices are then related by a correspondence (it is an involution again) called *isotomy*. The barycentric coordinates of points conjugate in isotomy are reciprocal.

The following theorem becomes clear if we observe that the projection of the centroid is again the centroid of the opposite face.

Theorem 2.2.16 *Let $P = (p_i)$ be an nb-point of an n-simplex Σ. Denote by P_{ij}, $i \neq j$, $i, j = 1, \ldots, n+1$, the projection of the point P from the $(n-2)$-dimensional space $\omega_i \cap \omega_j$ on the line A_iA_j, then by Q_{ij} the point symmetric to P_{ij} with respect to the midpoint of the edge A_iA_j. Then, for all i, j, $i \neq j$, the hyperplanes joining $\omega_i \cap \omega_j$ with the points Q_{ij} have a (unique) common point Q. This point is conjugate to P in the isotomy.*

We can also formulate dual notions and dual theorems. Instead of points, we study the position of hyperplanes with respect to the simplex. Recall that the dual barycentric coordinates of the hyperplane are formed by the $(n+1)$-tuple of coefficients, say $(\alpha_1, \ldots, \alpha_{n+1})_d$, of the hyperplane with the equation $\sum_k \alpha_k x_k = 0$. Thus the improper hyperplane has dual coordinates $(1, \ldots, 1)_d$, the $(n-1)$-dimensional face ω_1 dual coordinates $(1, 0, \ldots, 0)_d$, etc.

We can again define an *nb-hyperplane* with respect to Σ as a hyperplane not containing any vertex of Σ (thus with all dual coordinates different from zero). Four *nb*-hyperplanes $(\alpha_k)_d$, $(\beta_k)_d$, $(\gamma_k)_d$, and $(\delta_k)_d$ can be considered as forming a *dual quasiparallelogram* if for some $\rho \neq 0$

$$\frac{\alpha_k \, \gamma_k}{\beta_k \, \delta_k} = \rho, \qquad k = 1, \ldots, n+1.$$

Theorem 2.2.17 *The four nb-hyperplanes $(\alpha_k)_d$, $(\beta_k)_d$, $(\gamma_k)_d$, and $(\delta_k)_d$ form a quasiparallelogram with respect to Σ if and only if there exists in the pencil of quadrics $\lambda \sum_k \alpha_k x_k \sum_k \gamma_k x_k + \mu \sum_k \beta_k x_k \sum_k \delta_k x_k = 0$ a quadric which contains all vertices of Σ.*

Proof. This follows immediately from the fact that the quadric $\sum_{ik} g_{ik} x_i x_k = 0$ contains the vertex A_j if and only if $g_{jj} = 0$. □

We call two *nb*-hyperplanes α and γ *isotomically conjugate* if the four hyperplanes α, ν, γ, and ν, where ν is the improper hyperplane $\sum_k x_k = 0$, form a quasiparallelogram.

Theorem 2.2.18 *Two nb-hyperplanes are isotomically conjugate if and only if the pairs of their intersection points with any edge are formed by isotomically conjugate points on that edge, i.e. their midpoint coincides with the midpoint of the edge.*

The proof is left to the reader.

We should also mention that the *nb*-point (a_i) and the *nb*-hyperplane $\sum_k a_k x_k = 0$ are sometimes called *harmonically conjugate.* It is clear that we can formulate theorems such as the following: four *nb*-points form a quasiparallelogram if and only if the corresponding harmonically conjugate hyperplanes form a quasiparallelogram. It is also immediate that the centroid of Σ and the improper hyperplane are harmonically conjugate.

Before stating the next theorem, let us define as *complementary faces* of Σ the faces F_1 and F_2 of which F_1 is determined by some of the vertices of Σ and F_2 by all the remaining vertices. Observe that the distance between the two faces is equal to the distance between the mutually parallel hyperplanes H_1 and H_2, H_1 containing F_1 and H_2 containing F_2. In this notation, we shall prove:

Theorem 2.2.19 *Let the face F_1 be determined by the vertices A_j for $j \in J$, $J \subset N = \{1, \ldots, n+1\}$ so that the complementary face is determined by the vertices A_l for $l \in \bar{J} := N \setminus J$. Then the equation of H_1 is*

$$\sum_{j \in \bar{J}} x_j = 0,$$

and the equation of H_2 is

$$\sum_{l \in J} x_l = 0.$$

The distance $\rho(F_1, F_2)$ of both faces F_1 and F_2 is

$$\rho(F_1, F_2) = \frac{1}{\sqrt{\sum_{i,k \in J} q_{ik}}} \tag{2.12}$$

or, if $\rho^2(F_1, F_2) = z$, the number z is the only root of the equation

$$\det(M_0 - zC_J) = 0, \tag{2.13}$$

where M_0 is the matrix in (1.26) and $C_J = [c_{rs}]$, $r, s = 0, 1, \ldots, n+1$, $c_{ik} = 1$ if and only if $i \in J$ and $k \in \bar{J}$, or $i \in \bar{J}$ and $k \in J$, whereas $c_{rs} = 0$ in all other cases.

The common normal to both faces F_1 and F_2 joins the points $P_1 = (\delta_{iJ} \sum_{k \in J} q_{ik})$ and $P_2 = (\delta_{i\bar{J}} \sum_{k \in \bar{J}} q_{ik})$, the intersection points of this normal with F_1 and F_2; here, δ_{iJ} etc. is equal to one if $i \in J$, otherwise it is zero.

Proof. First of all, it is obvious that H_1 and H_2 have the properties that each contains the corresponding face and they are parallel to each other. The formula (2.12) then follows from the formula (1.43) applied to the distance of any vertex from the not-incident hyperplane. To prove the formula (2.13), observe that the determinant on the left-hand side can be written as

$$\det(M_0 + 2zP_0),$$

where P_0 is the matrix $[p_{rs}]$ with $p_{rs} = \delta_{rJ}\delta_{sJ}$ in the notation above. Since P_0 has rank one and M_0 is nonsingular with the inverse $-\frac{1}{2}Q_0$ by (1.29), the left-hand side of (2.13) is equal to

$$\det M_0 \det(I - zQ_0P_0).$$

Since the second determinant is $1 - z\sum_{i\in J, k\in J} q_{ik}$, the formula follows from the preceding.

The last assertion follows from (1.33) in Theorem 1.4.9. □

Remark 2.2.20 We shall call the minimum of the numbers $\rho(F_1, F_2)$ over all proper faces F_1 (the complementary face F_2 is then determined) the *thickness* $\tau(\Sigma)$ of the simplex Σ. It thus corresponds to the *minimal complete off-diagonal block* of the Gramian, so that

$$\tau(\Sigma) = \frac{1}{\sqrt{-\min_J \sum_{i\in J, j\notin J} q_{ij}}} \tag{2.14}$$

in the notation above. Observe that if the set of indices N can be decomposed in such a way that $N = N_1 \cup N_2$, $N_1 \cap N_2 = \emptyset$, where the indices i, j of all acute angles φ_{ij} belong to distinct N_is and the indices of all obtuse angles belong to the same N_i, then the thickness in (2.14) is realized for $J = N_1$. We shall call simplexes with this property *flat* simplexes. Observe also that the sum $\sum_{i\in J, k\in J} q_{ik}$ is equal to $\sum_{i\in \bar{J}, k\in \bar{J}} q_{ik}$ since the row sums of the matrix Q are equal to zero.

For the second part of this chapter, we shall investigate metric properties of quadrics (cf. Appendix, Section 5) in $\widehat{\mathcal{E}}_n$, in particular of the important *circumscribed Steiner ellipsoid* S of the n-simplex Σ. This ellipsoid S is the quadric whose equation in barycentric coordinates is

$$\sum_{i<k} x_i x_k = 0. \tag{2.15}$$

Let us start with a few general facts about quadrics in barycentric coordinates.

Let $\mathcal{B}$ be a nonsingular quadric with the equation

$$\sum_{i,k=1}^{n+1} b_{ik} x_i x_k = 0, \quad B = [b_{ik}] = B^T. \tag{2.16}$$

Such a quadric is called *central* if there is a proper point (the center), the polar of which is the improper hyperplane (1.13).

Lemma 2.2.21 *A nonsingular quadric (2.16) is central if and only if for the vector* $\mathbf{e}$ *of all ones*

$$e^T B^{-1} e \neq 0. \tag{2.17}$$

Proof. If the quadric in (2.16) is central and $C = (c_i)$ is the center, then

$$\sum_i c_i \neq 0 \tag{2.18}$$

and

$$\sum_{i,k} b_{ik}c_i x_k = 0$$

is the equation of the improper hyperplane, i.e.

$$\sum_i b_{ik}c_i = K, \text{ a nonzero constant, for all } k. \tag{2.19}$$

Thus in matrix form for $\hat{c} = [c_1, \ldots, c_{n+1}]^T$

$$B\hat{c} = Ke,$$

which implies

$$\sum_i c_i = e^T\hat{c} = Ke^T B^{-1}e.$$

By (2.18), we obtain (2.17).

Conversely, if (2.17) holds, then the unique solution c_i to

$$\sum_i b_{ik}c_i = 1,$$

which is $\hat{c} = B^{-1}e$, satisfies (2.18) and defines a center of (2.16). □

To find the axes of a nonsingular central quadric, let us use the following characteristic property of their improper points.

An improper point y is the improper point of an axis of a nonsingular central quadric if and only if the polar π of y is orthogonal to y, i.e. if and only if the orthogonal point to π coincides with y.

Theorem 2.2.22 *An improper point $y = (y_1, \ldots, y_{n+1})$ is the improper point of an axis for a nonsingular central quadric (2.16) if and only if the column vector $\hat{y} = [y_1, \ldots, y_{n+1}]^T$ is an eigenvector of the matrix QB corresponding to a nonzero eigenvalue λ of QB.*

The square l^2 of the length of the corresponding halfaxis is then

$$l^2 = -\frac{1}{\lambda e^T B^{-1} e}.$$

Remark 2.2.23 The square l^2 can even be negative (if (2.16) is not an ellipsoid).

Proof. By the mentioned characterization, y satisfying $\sum_i y_i = 0$ is the improper point of an axis of (2.16) if and only if the orthogonal point (by (1.33)) $z = (z_i)$, where

$$z_i = \sum_{j,k} q_{ij} b_{jk} y_k$$

to the polar hyperplane of y

$$\sum_{j,k} b_{jk} y_k x_j = 0$$

coincides with y, i.e. if and only if

$$z_i = \lambda y_i, \ i = 1, \ldots, n+1,$$

for some

$$\lambda \neq 0.$$

This yields for the column vector $\hat{y} \neq 0$ that

$$QB\hat{y} = \lambda \hat{y}, \ \lambda \neq 0. \tag{2.20}$$

The converse is also true.

Let us show that $\hat{y}^T B \hat{y} \neq 0$. Indeed, assume that $\hat{y}^T B \hat{y} = 0$. Denote by w the (nonzero) vector $w = B\hat{y}$, thus satisfying $\hat{y}^T w = 0$. By (2.20), $Qw = \lambda \hat{y}$ so that $w^T Q w = 0$. By Theorem A.1.40, w is a nonzero multiple of e; hence, $\hat{y} = \rho B^{-1} e$ and $\hat{y} w = 0$ implies $e^T B^{-1} e = 0$, a contradiction.

The length l of the corresponding halfaxis is the distance from the center c to the intersection point, say u, of the line cy with the quadric (strictly speaking, if it is real).

Thus, in a clear notation, let

$$u = c + \xi y \text{ for some } \xi, \tag{2.21}$$

where

$$\sum_{i,k} b_{ik} u_i u_k = 0.$$

This yields

$$\sum_{i,k} b_{ik} c_i c_k + 2\xi \sum_{i,k} b_{ik} c_i y_k + \xi^2 \sum b_{ik} y_i y_k = 0.$$

Since the middle term is zero by (2.19), we obtain

$$\xi^2 = -\frac{\hat{c}^T B \hat{c}}{\hat{y}^T B \hat{y}}.$$

We can assume that in the relation

$$B\hat{c} = Ke, \ K \text{ constant},$$

the number K and the vector $\hat{c}$ are such that $e^T \hat{c} = 1$, i.e. also

$$e^T B^{-1} e = \frac{1}{K}. \tag{2.22}$$

Then

$$e^T\hat{u} = 1$$

in (2.21) and the distance l between u and c satisfies by (1.9) the relation

$$\begin{aligned} l^2 &= -\frac{1}{2}\sum_{i,k} m_{ik}(u_i - c_i)(u_k - c_k) \\ &= -\frac{1}{2}\xi^2\sum_{i,k} m_{ik}y_iy_k \\ &= -\frac{1}{2}\xi^2\hat{y}^T M\hat{y} \\ &= -\frac{\xi^2}{2\lambda}\hat{y}^T MQB\hat{y} \end{aligned}$$

by (2.20).

Since

$$MQ = -2I_{n+1} - eq_0^T$$

by (1.21) and $\hat{y}^T e = 0$, we obtain

$$\hat{y}^T MQ = -2\hat{y}^T$$

and

$$\begin{aligned} l^2 &= \frac{\xi^2}{\lambda}\hat{y}^T B\hat{y} \\ &= -\frac{1}{\lambda}\hat{c}^T B\hat{c} \\ &= -\frac{K}{\lambda}\hat{c}^T e \\ &= -\frac{1}{\lambda e^T B^{-1} e} \end{aligned}$$

by (2.20) and (2.22). □

Theorem 2.2.24 *The center of the Steiner circumscribed ellipsoid S from (2.15) is the centroid $(1, \ldots, 1)$ (or $(\frac{1}{n+1}, \ldots, \frac{1}{n+1})$ in nonhomogeneous barycentric coordinates); S contains all the vertices A_i of Σ, and the tangent plane to S at A_i has equation*

$$\sum_{j\neq i} x_j = 0$$

and is parallel to ω_i. The squares l_i^2 of its halfaxes are given by

$$l_i^2 = \frac{n}{n+1}\frac{1}{\lambda_i};$$

here the λ_is are the nonzero eigenvalues of the matrix Q. The directions of the corresponding axes coincide with the eigenvectors $\hat{y} = (y_i)$ of Q. Also, the corresponding equations

$$\sum_{i=1}^{n+1} y_i x_i = 0$$

are the equations of the hyperplanes through the center of S orthogonal to the corresponding axes (hyperplanes of symmetry).

Proof. This follows immediately from Theorem 2.2.22 applied to $B = J - I$, $J = ee^T$ since

$$B^{-1} = \frac{1}{n}J - I,$$

$$QB = -Q,$$

and

$$e^T B^{-1} e = \frac{n+1}{n}.$$

It is easily seen that its center is the centroid $(1, \ldots, 1)$ (or $(\frac{1}{n+1}, \ldots, \frac{1}{n+1})$ in nonhomogeneous barycentric coordinates); S contains all the vertices A_i of Σ, the tangent plane to S at A_i has equation

$$\sum_{j \neq i} x_j = 0,$$

and is parallel to ω_i. □

Remark 2.2.25 Since the λ_is are positive, S is indeed an ellipsoid.

3

Qualitative properties of the angles in a simplex

3.1 Signed graph of a simplex

In this section, we intend to study interior (dihedral) angles in a simplex. In terms of the *quality* of an angle of a simplex, we shall understand that it belongs to just one of the following three classes: acute, obtuse, or right.

The relation (2.6) – together with the conditions of positive semidefiniteness – is a generalization of the theorem about the sum of angles in a triangle. For a triangle, it follows that at least two of the angles are acute. We intend to generalize this property for the n-simplex.

Theorem 3.1.1 *Suppose that φ_{ik}, $i, k = 1, \dots, n+1$, are the interior angles of an n-simplex Σ. Then there exists no nontrivial decomposition of the index set $N = \{1, \dots, n+1\}$ into two nonvoid subsets N_1, N_2 (i.e. $N_1 \cup N_2 = N$, $N_1 \cap N_2 = \emptyset$, $N_1 \neq \emptyset$, $N_2 \neq \emptyset$) such that $\varphi_{ik} \geq \frac{\pi}{2}$ for $i \in N_1$, $k \in N_2$.*

Proof. Suppose there exists such a decomposition. By (iii) of Theorem 1.3.3, the quadratic form

$$\sum_{i \in N} x_i^2 - \sum_{i,k \in N, i \neq k} x_i x_k \cos \varphi_{ik} \geq 0 \tag{3.1}$$

(it is positive semidefinite) and equality is attained if and only if $x_i = \rho p_i$, $p_i > 0$. By Theorem A.1.40, then also

$$p_i - \sum_{k \in N, k \neq i} p_k \cos \varphi_{ik} = 0, \qquad i \in N. \tag{3.2}$$

Multiply the ith relation (3.2) for $i \in N_1$ by p_i and add for $i \in N_1$. We obtain

$$\sum_{i \in N_1} p_i^2 - \sum_{i,k \in N_1, i \neq k} p_i p_k \cos \varphi_{ik} - \sum_{i \in N_1, k \in N_2} p_i p_k \cos \varphi_{ik} = 0. \tag{3.3}$$

Since $\cos \varphi_{ik} \leq 0$ for $i \in N_1$, $k \in N_2$, the last summand in (3.3) is nonnegative. The sum of the remaining two summands is also nonnegative

$$\sum_{i \in N_1} p_i^2 - \sum_{i,k \in N_1, i \neq k} p_i p_k \cos \varphi_{ik} \geq 0; \tag{3.4}$$

this follows by substitution of

$$x_i = p_i \ \text{ for } \ i \in N_1, \quad x_j = 0 \ \text{ for } \ j \in N_2 \tag{3.5}$$

into (3.1).

Because of (3.3), there must be equality in (3.4). However, that contradicts the fact that equality in (3.1) is attained only for $x_i = \rho p_i$ and not for the vector in (3.5) (observe that $N_2 \neq \emptyset$). □

To better visualize this result, we introduce the notion of the *signed graph* G_Σ of the n-simplex Σ (cf. Appendix).

Definition 3.1.2 Let Σ be an n-simplex with $(n-1)$-dimensional faces $\omega_1, \ldots, \omega_{n+1}$. Denote by G^+ the undirected graph with $n+1$ nodes $1, 2, \ldots, n+1$, and those edges (i,k), $i \neq k$, $i,k = 1, \ldots, n+1$, for which the interior angle φ_{ik} of the faces ω_i, ω_k is acute

$$\varphi_{ik} < \frac{\pi}{2}.$$

Analogously, denote by G^- the graph with the same set of nodes but with those edges (i,k), $i \neq k$, $i,k = 1, \ldots, n+1$, for which the angle φ_{ik} is obtuse

$$\varphi_{ik} > \frac{\pi}{2}.$$

We can then consider G^+ and G^- as the positive and negative parts of the *signed graph* G_Σ of the simplex Σ. Its nodes are the numbers $1, 2, \ldots, n+1$, its positive edges are those from G^+, and its negative edges are those from G^-.

Now we are able to formulate and prove the following theorem ([4], [7]).

Theorem 3.1.3 *If G_Σ is the signed graph of an n-simplex Σ, then its positive part is a connected graph.*

Conversely, if G is an undirected signed graph (i.e. each of its edges is assigned a sign + or −) with $n+1$ nodes $1, 2, \ldots, n+1$ such that its positive part is a connected graph, then there exists an n-simplex Σ, whose graph G_Σ is G.

Proof. Suppose first that the positive part G^+ of the graph G_Σ of some n-simplex Σ is not connected. Denote by N_1 the set consisting of the index 1 and of all such indices k that there exists a sequence of indices $j_0 = 1, j_1, \ldots, j_t = k$ such that all edges (j_{s-1}, j_s) for $s = 1, \ldots, t$ belong to G^+. Further, let N_2 be the set of all the remaining indices. Since G^+ is not connected, $N_2 \neq \emptyset$, and the following holds: whenever $i \in N_1$, $k \in N_2$, then $\varphi_{ik} \geq \frac{1}{2}\pi$ (otherwise $k \in N_1$). That contradicts Theorem 3.1.1.

To prove the converse part, observe that the quadratic form

$$f = \sum_{(i,j)\in G^+} (\xi_i - \xi_j)^2$$

is positive semidefinite and attains the value zero only for $\xi_1 = \xi_2 = \cdots = \xi_{n+1}$; this follows easily from the connectedness of the graph G^+. Therefore, the principal minors of order $1, \ldots, n$ of the corresponding $(n+1) \times (n+1)$ matrix are strictly positive (and the determinant is zero). Thus there exists a sufficiently small positive number ε so that also the form

$$f_1 \equiv f - \varepsilon \sum_{(i,j)\in G^-} (\xi_i - \xi_j)^2 \equiv \sum_{i,j=1} v_{ij}\xi_i\xi_j$$

is positive semidefinite and equal to zero only for $\xi_1 = \xi_2 = \cdots = \xi_{n+1}$. By Theorem 1.3.3, there exists an n-simplex Σ, the interior angles φ_{ij} of which satisfy the conditions

$$\cos\varphi_{ij} = -\frac{v_{ij}}{\sqrt{v_{ii}}\sqrt{v_{jj}}}.$$

Since

$$\begin{aligned} \varphi_{ij} &< \tfrac{\pi}{2} \quad \text{for } (i,j) \in G^+, \\ \varphi_{ij} &> \tfrac{\pi}{2} \quad \text{for } (i,j) \in G^-, \\ \text{and} \qquad \varphi_{ij} &= \tfrac{\pi}{2} \quad \text{for the remaining } (i,j), i \neq j, \end{aligned}$$

Σ fulfills the conditions of the assertion. □

It is well known that a connected graph with $n+1$ nodes has at least n edges (cf. Appendix, Theorem A.2.4), and there exist connected graphs with $n+1$ nodes and n edges (so-called *trees*). Thus we obtain:

Theorem 3.1.4 *Every n-simplex has at least n acute interior angles. There exist n-simplexes which have exactly n acute interior angles.*

Remark 3.1.5 The remaining $\binom{n}{2}$ angles can be either obtuse or right.

This leads us to the following definition.

Definition 3.1.6 An n-simplex which has n acute interior angles and all the remaining $\binom{n}{2}$ interior angles right will be called a *right simplex.*

Corollary 3.1.7 *The graph of a right simplex is a tree.*

We shall return to this topic later in Chapter 4, Section 1.

There is another, perhaps more convenient, approach to visualizing the angle properties of an n-simplex. It is based on the fact that the interior angle φ_{ij} has the opposite edge A_iA_j in the usual notation.

Definition 3.1.8 Color the edge A_iA_j *red* if the opposite interior angle φ_{ij} is acute, and color it *blue* if φ_{ij} is obtuse.

The result of Theorem 3.1.3 can thus be formulated as follows:

Theorem 3.1.9 *The coloring of every simplex has the property that the red edges connect all vertices of the simplex. If we color some edges of a simplex in red, some in blue, and leave some uncolored, but in such a way that the red edges connect the set of all vertices, then there exists a "deformation" of the simplex for which the opposite interior angles to red edges are acute, the opposite interior angles to blue edges are obtuse, and the opposite interior angles to uncolored edges are right.*

Example 3.1.10 Let the points $A_1, \ldots, A_{n+1}$ in a Euclidean n-dimensional space $\mathcal{E}_n$ be given using the usual Cartesian coordinates

$$\begin{aligned} A_1 &= (0, 0, \ldots, 0), \\ A_2 &= (a_1, 0, \ldots, 0), \\ A_3 &= (a_1, a_2, 0, \ldots, 0), \\ A_4 &= (a_1, a_2, a_3, 0, \ldots, 0), \\ &\ldots \\ A_{n+1} &= (a_1, a_2, a_3, \ldots, a_n), \end{aligned}$$

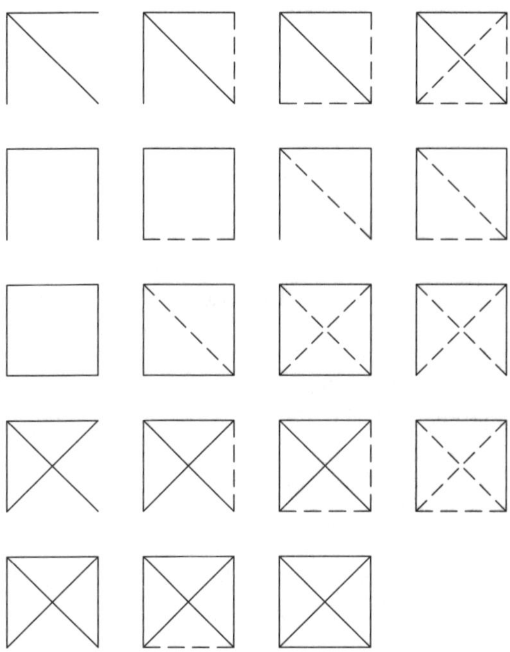

Fig. 3.1

where $a_1, a_2, \ldots, a_n$ are some positive numbers. These points are linearly independent, thus forming vertices of an n-simplex Σ. The $(n-1)$-dimensional faces $\omega_1, \omega_2, \ldots, \omega_{n+1}$ are easily seen to have equations $\omega_1 \equiv x_1 - a_1 = 0$, $\omega_2 \equiv a_2 x_1 - a_1 x_2 = 0$, $\omega_3 \equiv a_3 x_2 - a_2 x_3 = 0, \ldots, \omega_{n+1} \equiv x_n = 0$. Using the formula (1.2), we see easily that all pairs ω_i, ω_j, $i \neq j$, are perpendicular except the pairs (ω_1, ω_2), $(\omega_2, \omega_3), \ldots,$ (ω_n, ω_{n+1}). Therefore, only the edges $A_1A_2, A_2A_3, \ldots, A_nA_{n+1}$ are colored (in red), and the graph of Σ is a path. Observe that these edges are mutually perpendicular. In addition, all two-dimensional faces $A_iA_jA_k$ are right triangles with the right angle at A_j if $i < j < k$. Thus also the midpoint of A_1A_{n+1} is the center of the circumscribed hypersphere of Σ because of the Thalet theorem.

Example 3.1.11 In Fig. 3.1, all possible colored graphs of a tetrahedron are depicted. The red edges are drawn unbroken, the blue edges dashed. The missing edges correspond to right angles.

3.2 Signed graphs of the faces of a simplex

In this section, we shall investigate some further properties of the interior angles in an n-simplex and its smaller-dimensional faces. In particular, we shall, as in Chapter 1, be interested in whether these angles are acute, right, or obtuse.

For this purpose, we shall use Theorem 1.3.3 and the Gramian Q of the simplex Σ. By Theorem 2.1.1, the following is immediate.

Corollary 3.2.1 *The signed graph of the simplex Σ is (if the vertices and the nodes are numbered in the same way) the negative of the signed graph of the Gramian Q of Σ.*

Remark 3.2.2 The *negative* of a signed graph is the graph with the same edges in which the signs are changed to the opposite.

Our task will now be to study the faces of the n simplex Σ, using the results of Chapter 1, in particular the formula (1.21). First of all, we would like to find the Menger matrix and the Gramian of such a face. Thus let Σ be an n-simplex, and Σ' be its face determined by the first $m+1$ vertices. Partition the matrices in (1.21). We then have that

$$\begin{bmatrix} 0 & e_1^T & e_2^T \\ e_1 & M_{11} & M_{12} \\ e_2 & M_{21} & M_{22} \end{bmatrix} \begin{bmatrix} q_{00} & q_{01}^T & q_{02}^T \\ q_{01} & Q_{11} & Q_{12} \\ q_{02} & Q_{21} & Q_{22} \end{bmatrix} = -2I_{n+2}, \tag{3.6}$$

where M_{11}, Q_{11} are $(m+1) \times (m+1)$ matrices corresponding to the vertices in Σ', etc. It is clear that M_{11} is the Menger matrix of Σ'.

To obtain the Gramian of Σ', we have, by the formula (1.21), to express in terms of the matrices assigned to Σ the extended Gramian

$$\hat{Q}_0 = \begin{bmatrix} \hat{q}_{00} & \hat{q}_{01}^T \\ \hat{q}_{01} & \hat{Q}_{11} \end{bmatrix}.$$

By the formula (1.29), this matrix is the $(-\frac{1}{2})$-multiple of the inverse of the extended Menger matrix

$$\begin{bmatrix} 0 & e_1^T \\ e_1 & M_{11} \end{bmatrix}.$$

Using the formula (A.5), we obtain

$$\begin{bmatrix} 0 & e_1^T \\ e_1 & M_{11} \end{bmatrix}^{-1} = \left(-\frac{1}{2}\right) \begin{bmatrix} q_{00} & q_{01}^T \\ q_{01} & Q_{11} \end{bmatrix} - \left(-\frac{1}{2}\right) \begin{bmatrix} q_{02}^T \\ Q_{12} \end{bmatrix} Q_{22}^{-1} [q_{02} \ \ Q_{21}],$$

i.e.

$$\begin{bmatrix} 0 & e_1^T \\ e_1 & M_{11} \end{bmatrix}^{-1} = \left(-\frac{1}{2}\right) \begin{bmatrix} q_{00} - q_{02}^T Q_{22}^{-1} q_{02} & q_{01}^T - q_{02}^T Q_{22}^{-1} Q_{21} \\ q_{01} - Q_{12} Q_{22}^{-1} q_{02} & Q_{11} - Q_{12} Q_{22}^{-1} Q_{21} \end{bmatrix}. \tag{3.7}$$

This means that the Gramian corresponding to the m-simplex Σ' is the Schur complement (Appendix (A.4)) $Q_{11} - Q_{12}Q_{22}^{-1}Q_{21}$ of Q_{22} in the Gramian of Σ.

Let us summarize, having in mind that the choice of the numbering of the vertices is irrelevant:

Theorem 3.2.3 *Let Σ be an n-simplex with vertices A_i, $i \in N = \{1, \dots, n+1\}$. Denote by Σ' the face of Σ determined by the vertices A_j for $j \in M = \{1, \dots, m+1\}$ for some $m < n$. If the extended Menger matrix of Σ is partitioned as in* (3.6)*, then the extended Gramian of Σ'*

$$\hat{Q}_0 = \begin{bmatrix} \hat{q}_{00} & \hat{q}_{01}^T \\ \hat{q}_{01} & \hat{Q}_{11} \end{bmatrix}$$

is equal to

$$\begin{bmatrix} q_{00} - q_{02}^T Q_{22}^{-1} q_{02} & q_{01}^T - q_{02}^T Q_{22}^{-1} Q_{21} \\ q_{01} - Q_{12} Q_{22}^{-1} q_{02} & Q_{11} - Q_{12} Q_{22}^{-1} Q_{21} \end{bmatrix}.$$

In particular, the Gramian of Σ' is the Schur complement of the Gramian of Σ with respect to the indices in $N \backslash M$.

In Chapter 1, Section 3, we were interested in qualitative properties of the interior angles in a simplex; under *quality* of the angle, we understood the property of being acute, obtuse, or right. Using the result of Theorem 3.2.3, we can ask if something analogous can be proved for the faces.

In view of Corollary 3.2.1, it depends on the signed graph of the Schur complement of the Gramian of the simplex.

In the graph-theoretical approach to the problem of how the zero–nonzero structure of the matrix of a system of linear equations is changed by elimination of one (or, more generally, several) unknown, the notion of the elimination graph was discovered. Also, elimination of a group of unknowns can be performed (under mild existence conditions) by performing a sequence of simple eliminations where just one unknown is eliminated at a time. The theory is based on the fact (Appendix (A.7)) that the Schur complement of the Schur complement is the Schur complement again. In our case, the situation is even more complicated by the fact that we have to consider the signs of the entries. This means that only rarely can we expect definite results in this respect.

However, there is one class of simplexes for which this can be done completely. This class will be studied in the next section.

3.3 Hyperacute simplexes

In this section, we investigate n-simplexes, *no* interior angle of which (between $(n-1)$-dimensional faces) is obtuse. We call these simplexes *hyperacute.* For completeness, we consider all 1-simplexes as hyperacute as well. In addition, we say that a simplex is *strictly hyperacute* if *all* its interior angles are acute.

As we have seen in Section 3.1, the signed graph of such a simplex has positive edges only and its colored graph does not contain a blue edge. Its Gramian Q from (1.21) is thus a singular M-matrix (see Appendix (A.3.9)).

The first, and most important result, is the following:

Theorem 3.3.1 *Every face of a hyperacute simplex is also a hyperacute simplex. In addition, the graph of the face is uniquely determined by the graph of the original simplex and is obtained as the elimination graph after removing the graph nodes corresponding to simplex vertices not contained in the face.*

More explicitly, using the coloring of the simplex (this time, of course, without the blue color), the following holds:

Suppose that the edges of the simplex are colored as in Theorem 3.1.9. Then an edge (i,k), $i \neq k$, in the face obtained by removing the vertex set S will be colored in red if and only if there exists a path from i to k which uses red edges only, all of which vertices (different from i and k) belong to S. Otherwise, the edge (i,k) will remain uncolored.

Proof. It is advantageous to use matrices. Let Σ be an n-simplex, and Σ' be its face determined by the first $m+1$ vertices. Partition the matrices as in (3.6). We saw in (3.7) that the Gramian corresponding to the m-simplex Σ' is the Schur complement $Q_{11} - Q_{12}Q_{22}^{-1}Q_{21}$ of Q_{11} in the Gramian of Σ. This Schur complement is by Theorem A.3.8 again a (singular) M-matrix with the annihilating vector of all ones so that Σ' is a hyperacute simplex.

As shown in Theorem A.3.5, the graph of the Schur complement is the elimination graph obtained by elimination of the rows and columns as described, so that the last part follows. □

Since every elimination graph of a complete graph is also complete, we obtain the following result.

Theorem 3.3.2 *Every face of a strictly hyperacute simplex is again a strictly hyperacute simplex.*

3.4 Position of the circumcenter of a simplex

In this section, we investigate how the *quality* of interior angles of the simplex relates to the position of the circumcenter. The case $n = 2$ shows such a relationship exists: in a strictly acute triangle, the circumcenter is an interior point of the triangle; the circumcenter of a right triangle is on the boundary; and in the obtuse triangle, it is an exterior point (in the obtuse angle). We generalize these properties only in a qualitative way, i.e. with respect to the halfspaces determined by the $(n-1)$-dimensional faces. It turns out that this relationship is well characterized by the use of the *extended graph* of the simplex (cf. [10]). Here is the definition:

Definition 3.4.1 Denote by $A_1, \ldots, A_{n+1}$ the vertices, and by $\omega_1, \ldots, \omega_{n+1}$ the $(n-1)$-dimensional faces (ω_i opposite A_i), of an n-simplex Σ. Let C be the circumcenter. The extended graph G^*_Σ is obtained by extending the usual graph G_Σ by one more node 0 (zero) corresponding to the circumcenter C as follows: the node 0 is connected with the node k, $1 \le k \le n+1$ (corresponding to ω_k) by an edge if and only if ω_k does not contain C; this edge is positive (respectively, negative) if C is in the same (respectively, the opposite) halfspace determined by ω_k as the vertex A_k.

In Fig. 3.2 a, b, c, the extended graphs of an acute, right, and obtuse triangle are depicted. The positive edges are drawn unbroken, the obtuse dashed. The right or obtuse angle is always at vertex A_2.

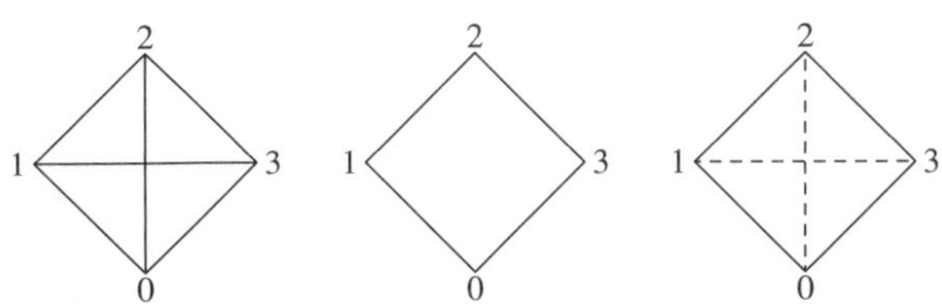

Fig. 3.2

By the definition of the extended graph, edges ending in the additional node 0 correspond to the signs of the (inhomogeneous) barycentric coordinates of the circumcenter C. If the kth coordinate is zero, there is no edge $(0,k)$ in G^*_Σ; if it is positive (respectively, negative), the edge $(0,k)$ is positive (respectively, negative). By Theorem 2.1.1, this means:

Theorem 3.4.2 *The extended graph G^*_Σ of the n-simplex Σ is the negative of the signed graph of the $(n+2)\times(n+2)$ matrix $[q_{rs}]$, the extended Gramian of this simplex, in which the distinguished vertex 0 corresponds to the first row (and column).*

Remark 3.4.3 As in Remark 3.2.2, the negative of a signed graph is the graph with the same edges in which the signs are changed to the opposite.

The proof of Theorem 3.4.2 follows from the formulae in Theorem 2.1.1 and the definitions of the usual and extended graphs.

In Theorem 3.1.3 we characterized the (usual) graphs of n-simplexes, i.e. we found necessary and sufficient conditions for a signed graph to be a graph of some n-simplex. Although we shall not succeed in characterizing extended graphs of simplexes in a similar manner, we find some interesting properties of these. First of all, we show that the exclusiveness of the node 0 is superfluous.

Theorem 3.4.4 *Suppose a signed graph Γ on $n+2$ nodes is an extended graph of an n-simplex Σ_1, the node u_1 of Γ being the distinguished node corresponding to the circumcenter C_1 of Σ_1. Let u_2 be another node of Γ. Then there exists an n-simplex Σ_2, the extended graph of which is also Γ, and such that u_2 is the distinguished node corresponding to the circumcenter C_2 of Σ_2.*

Proof. Let $[q_{rs}]$ be the Gramian of the simplex Σ_1. This means that, for an appropriate numbering of the vertices of the graph Γ, in which u_1 corresponds to index 0, we have for $r \neq s$, $r,s = 0,\dots,n+1$,

$$(r,s) \text{ is a } \left\{ \begin{array}{c} \text{positive} \\ \text{negative} \end{array} \right\} \text{ edge of } \Gamma \text{ if and only if } \left\{ \begin{array}{c} q_{rs} < 0 \\ q_{rs} > 0 \end{array} \right\}.$$

We can assume that the vertex u_2 corresponds to the index $n+1$. The matrix $-2[q_{rs}]^{-1}$ equals the matrix $[m_{rs}]$, which satisfies the conditions of Theorem 1.2.4. We show that the matrix $[m'_{rs}]$ with entries

$$\begin{aligned}
m'_{rr} &= 0 && (r = 0,\dots,n+1), \\
m'_{i0} &= m'_{0i} = 1 && (i = 1,\dots,n+1), \\
m'_{\alpha,n+1} &= m'_{n+1,\alpha} = \frac{1}{m_{\alpha,n+1}} && (\alpha = 1,\dots,n), \\
m'_{\alpha\beta} &= \frac{m_{\alpha\beta}}{m_{\alpha,n+1}m_{\beta,n+1}} && (\alpha,\beta = 1,\dots,n)
\end{aligned} \tag{3.8}$$

also fulfills the condition of Theorem 1.2.4

$$\sum_{i,k=1}^{n+1} m'_{ik}x'_ix'_k < 0, \qquad \text{when} \qquad \sum_1^{n+1} x'_i = 0, \ (x'_i) \neq 0.$$

Thus suppose $(x'_i) \neq 0$, $\sum_1^{n+1} x'_i = 0$. Define the numbers $x_\alpha = \dfrac{x'_\alpha}{m_{\alpha,n+1}}$, $\alpha = 1, \dots, n$, $x_{n+1} = -\sum_{\alpha=1}^{n} x_\alpha$. We have then

$$\begin{aligned}
\sum m'_{ik}x'_ix'_k &= \sum_{\alpha,\beta=1}^{n} m_{\alpha\beta}\frac{x'_\alpha}{m_{\alpha,n+1}}\frac{x'_\beta}{m_{\beta,n+1}} + 2x'_{n+1}\sum_{\alpha=1}^{n}\frac{x'_\alpha}{m_{\alpha,n+1}} \\
&= \sum_{\alpha,\beta=1}^{n} m_{\alpha\beta}x_\alpha x_\beta - 2\sum_{\alpha=1}^{n} x'_\alpha \sum_{\alpha=1}^{n} x_\alpha \\
&= \sum_{\alpha,\beta=1}^{n} m_{\alpha\beta}x_\alpha x_\beta + 2x_{n+1}\sum_{\alpha=1}^{n} m_{\alpha,n+1}x_\alpha \\
&= \sum_{i,k=1}^{n+1} m_{ik}x_ix_k < 0
\end{aligned}$$

by Theorem 1.2.4. This means that there exists an n-simplex Σ_2, the matrix of which is $[m'_{rs}]$. However, the matrix $[m'_{rs}]$ arises from $[m_{rs}]$ by multiplication from the right and from the left by the diagonal matrix $D = \operatorname{diag}(d_r)$, where

$$\begin{aligned}
d_\alpha &= \frac{1}{m_{\alpha,n+1}}, \qquad \alpha = 1, \dots, n, \\
d_0 &= d_{n+1} = 1,
\end{aligned}$$

and by exchanging the first and the last row and column. It follows that the inverse matrix $-\frac{1}{2}[q'_{rs}]$ to the matrix $[m'_{rs}]$ arises from the matrix $-\frac{1}{2}[q_{rs}]$ by multiplication by the matrix D^{-1} from both sides and by exchanging the first and last row and column. Since the matrix D^{-1} has positive diagonal entries, the signs of the entries do not change so that Γ will again be the extended graph of the n-simplex Σ_2. The exchange, however, will lead to the fact that the node u_2 will be distinguished, corresponding to the circumcenter of Σ_2. □

Remark 3.4.5 The transformation (3.8) corresponds to the spherical inversion, which transforms the vertices of the simplex Σ_1 into vertices of the simplex Σ_2. The center of the inversion is the vertex A_{n+1}. This also explains the geometric meaning of the transformation already mentioned in Remark 1.4.5.

Theorem 3.4.6 *If we remove from the extended graph of an n-simplex an arbitrary node, then the positive part of the resulting graph is connected.*

Proof. This follows from Theorems 3.1.3 and 3.4.4. □

Theorem 3.4.7 *The node connectivity number of the positive part of the extended graph with at least four nodes is always at least two.*[1]

Proof. This is an immediate consequence of the previous theorem. □

Let us return to Theorem 3.4.2. We can show[2] that the following theorem holds.

Theorem 3.4.8 *The set of extended graphs of n-simplexes coincides with the set of all negatively taken signed graphs of real nonsingular symmetric matrices of degree $n+2$, all principal minors of order $n+1$ which are equal to zero, have signature n, and the annihilating vector of one arbitrary principal submatrix of degree $n+1$ is positive. Exactly these matrices are the Gramians $[q_{rs}]$ of n-simplexes.*

The following theorem, the proof of which we omit (see [10], Theorem 3,12), expresses the nonhomogeneous barycentric coordinates of the circumcenter by means of the numbers q_{ij}, i.e. essentially by means of the interior angles of the n-simplex.

Theorem 3.4.9 *Suppose $[q_{ij}]$, $i, j = 1, \dots, n+1$, is the matrix Q corresponding to the n-simplex Σ. Then the nonhomogeneous barycentric coordinates c_i of the circumcenter of Σ can be expressed by the formulae*

$$c_i = \rho \sum_S (2 - \sigma_i(S))\pi(S), \tag{3.9}$$

where the summation is extended over all spanning trees S of the graph G_Σ

$$\pi(S) = \prod_{(p,q)\in E} (-q_{pq}),$$

$\sigma_i(S)$ is the degree of the node i in the spanning tree $S = (N, E)$ (i.e. the number of edges from S incident with i), and

$$\rho = \frac{1}{2\sum_S \pi(S)}.$$

This implies the following corollary.

[1] The node connectivity of a connected graph G is k, if the graph obtained by deleting any $k-1$ nodes is still connected, but after deleting some k nodes becomes disconnected (or void).

[2] See [10], where, in fact, all the results of this chapter are contained.

Theorem 3.4.10 *If we remove in the extended graph G^* of any n-simplex one node k, then the resulting graph G_k has the following properties:*

(i) *If j is a node with degree 2 in G_k, which is also a cut-node in G_k, then (j,k) is not an edge in G^*.*

(ii) *Every node with degree 1 in G_k is joined by a positive edge with k in G^*. If, in addition, G_k is a positive graph, then:*

(iii) *every node with degree 2 in G_k, which is not a cut-node in G_k, is joined with k by a positive edge in G^*.*

(iv) *Every node in G_k with degree at least 3, which is a cut-node in G_k, is joined with k by a negative edge in G^*.*

Proof. By Theorem 3.4.6, there exists a simplex whose extended graph is G^* with node k corresponding to the circumcenter. Thus G_k is its usual graph. Let us use the formula (3.9) from Theorem 3.4.9 and observe that $\rho > 0$. The assertion (i) follows from this formula since every cut-node j with degree 2 in G_k has degree 2 in every spanning tree of G_k, so that $\sigma_j(K) = 2$ and $c_j = 0$. The assertion 2 follows from Theorem 3.4.7 since a node in G^* cannot have degree 1.

Suppose now that G_k is a positive graph. If a node j has degree 2 and is not a cut-node in G_k, then every summand in (3.9) is nonnegative; however, there exists at least one positive summand since for some spanning tree in G_k, j has degree 1.

The assertion (iv) follows also from (3.9) since $\sigma_j(S) \geq 2$, whereas for some spanning tree S, $\sigma_j(S) > 2$. □

Let us consider now so-called *totally hyperacute simplexes.*

Definition 3.4.11 An n-simplex ($n \geq 2$) is called *totally hyperacute* if it is hyperacute and if its circumcenter is either an interior point of the simplex, or an interior point of one of its faces.

Remark 3.4.12 This means that the extended Gramian $[q_{rs}]$ from (1.27) has all off-diagonal entries nonpositive. A simplex whose circumcenter is an interior point is sometimes called *well-centered.*

Theorem 3.4.13 *Every m-dimensional face ($2 \leq m < n$) of a totally hyperacute n-simplex is again a totally hyperacute simplex. The extended graph of this face Σ_1 is by the extended graph of the given simplex uniquely determined and is obtained as its elimination graph by eliminating the vertices of the graph which correspond to all the $(n-1)$-dimensional faces containing Σ_1.*

Proof. Suppose Σ is a totally hyperacute n-simplex with vertices A_i and $(n-1)$-dimensional faces ω_i, $i = 1, \ldots, n+1$. Since the Gauss elimination operation is transitive, it suffices to prove the theorem for the case $m = n-1$. Thus let Σ_1 be the $(n-1)$-dimensional face in ω_{n+1}. If $\hat{M} = [m_{rs}]$ and $\hat{Q} = [q_{rs}]$

are the matrices of Σ from Corollary 1.4.3, so that $\hat{M}\,\hat{Q} = -2I$, I being the identity matrix, $r, s = 0, 1, \ldots, n+1$, then the analogous matrices for Σ_1 are $\tilde{M} = [m_{r's'}]$, $\tilde{Q} = [\tilde{q}_{r's'}]$, $r', s' = 0, 1, \ldots, n$, where

$$\tilde{q}_{r's'} = q_{r's'} - \frac{q_{r',n+1}q_{s',n+1}}{q_{n+1,n+1}}. \tag{3.10}$$

Indeed, these numbers fulfill the relation $\tilde{M}\,\tilde{Q} = -2\tilde{I}$, where $\tilde{I}$ is the identity matrix of order $n+1$. Since $q_{n+1,n+1} > 0$ and $q_{r's'} \le 0$ for $r' \ne s'$, $r', s' = 0, 1, \ldots, n$, we obtain by (3.10) that $\tilde{q}_{r's'} \le 0$.

Therefore, Σ_1 is also a totally hyperacute simplex. The extended graph of Σ_1 is a graph with nodes $0, 1, \ldots, n$. Its nodes r' and s' are joined by a positive edge if and only if $\tilde{q}_{r's'} < 0$, i.e. by (3.10) if and only if at least one of the following cases occurs:

(i) $q_{r's'} < 0$,
(ii) both inequalities $q_{r',n+1} < 0$ and $q_{s',n+1} < 0$ hold.

Analogous to the proof of Theorem 3.3.1, this means that $Q^*_{\Sigma_1}$ is the elimination graph of Q^*_{Σ} obtained by elimination of the node $\{n+1\}$. □

Theorem 3.4.14 *A positive polygon (circuit) is the extended signed graph of a simplex, namely of the simplex in Example 3.1.10.*

Proof. Evident from the results in the example. □

Combining the two last theorems, we obtain:

Theorem 3.4.15 *The extended graph of a totally hyperacute simplex is either a polygon (circuit), or a graph with node connectivity at least* 3.

Proof. By Theorem 3.4.7, the node connectivity of such a graph is at least 2. Suppose it is equal to 2 and that the number of its nodes is at least 4. Suppose that two of its nodes i and j form a cut. We show that the degree of the node i is 2. This will easily imply that every neighboring node to i also has degree 2 and forms a cut with j, or is a neighbor with j. By connectivity, the graph is then a polygon (circuit).

To prove the assertion, we use (iv) of Theorem 3.4.10 for the node j. Since the node i is then a cut-node in G_j, it cannot have degree ≥ 3 (since there is no negative edge in the graph), and thus also not degree 1. It has thus degree 2 and the proof is complete. □

Let us mention some consequences of this theorem.

Theorem 3.4.16 *A positive graph with a node of degree* 2 *is an extended graph of a simplex if and only if it is a polygon (circuit).*

Proof. Evident. □

Theorem 3.4.17 *Two-dimensional faces of a totally hyperacute simplex are either all right, or all acute triangles.*

Proof. If the node connectivity of the extended graph of a totally hyperacute simplex is at least 3, then every elimination graph with four vertices has node connectivity 3. By Theorem 3.4.13, every two-dimensional face is an acute triangle.

The case that the vertex connectivity is 2 follows from Theorem 4.1.8, which will be (independently) proved in Chapter 4, section 1. □

The next theorem presents a very strong property of extended graphs of totally hyperacute simplexes. We might conjecture that it even characterizes these graphs, i.e. that this condition is also sufficient.

Theorem 3.4.18 *Suppose that G_0 is the extended graph of a totally hyperacute n-simplex. If we remove from G_0 a set of k nodes and all incident edges, then the resulting graph G_1 has at most k components. If it has exactly k components, then G_0 is either a polygon (circuit), or $n+2=2k$ and there is no edge in G_0 joining any two removed nodes from G_0 or any two nodes in G_1.*

Proof. Denote by $N=\{0,1,\dots,n+1\}$ the set of nodes of the graph G_0. Let us remove from N a subset N_1 containing $k\geq 1$ nodes. We shall use induction. The theorem is correct for $k=1$. Suppose $k\geq 2$ and that the node 0 belongs to N_1, which is possible by Theorem 3.4.4. Denote $N_2=N_1\setminus\{0\}$. By Theorem 3.4.9, there exist numbers q_{ij}, $i,j=1,\dots,n+1$, and numbers c_i, $i=1,\dots,n+1$, such that

$$c_i=\rho\sum_S(2-\sigma_i(S))\pi(S),$$

where we sum over all spanning trees S of the graph G obtained by removing from G_0 the node 0 and edges incident with 0, where $\sigma_i(S)$ is the degree of the node i in S, and

$$\pi(S)=\prod_{(pq)\in S}(-q_{pq}).$$

All the numbers $\pi(S)$ are positive since $q_{pq}<0$ for $p\neq q$, so the number ρ is also positive. Denote by l the number of components of G_1 and by S an arbitrary spanning tree of G. Let $e(S)$ be the number of edges of S between nodes in N_2, and $l(S)$ the number of components obtained after removing from S the nodes in N_2 (and the incident edges). A simple calculation yields

$$\sum_{i\in N_2}\sigma_i(S)=n_2l(S)+e(S)-1,$$

where $n_2 = k - 1$ is the number of nodes in N_2. Since $c_i \geq 0$ and $l(S) \geq l$

$$\begin{aligned} 0 \leq \sum_{i \in N_2} s_i &= \sum_S \left[\sum_{i \in N_2} (2 - \sigma_i(S)) \right] \pi(S) \\ &= \sum_S [2(k-1) - (k-1) - l(S) - e(S) + 1] \pi(S) \\ &\leq \sum_S (k - l(S) - e(S)) \pi(S) \\ &\leq \sum_S (k - l) \pi(S). \end{aligned}$$

Thus $k \geq l$, which means that the number of components of the graph G_1 is at most k. In the case that $k = l$, $c_i = 0$ for $i \in N_2$, and further $l(S) = l$, and $e(S) = 0$ for all spanning trees S, so that indeed there is no edge in G_0 between two nodes from N_1. Suppose that G_0 is not a polygon (circuit). Then the node connectivity of G_0 is, by Theorem 3.4.15, at least 3 and the graph G has no cut-node.

Suppose that some component of G_1 has at least two nodes u_1, u_2. Since there is no cut-node in G, we can find for every node $v \in N_2$ a path $u_1 \ldots v \ldots u_2$ in G. This path can be completed into a spanning tree S_0 of G. However, u_1 and u_2 are in different components of the graph obtained from S_0 by removing the nodes from N_2, i.e. $l(S_0) \geq l + 1$. This contradiction with $l(S) = l$ for all spanning trees S proves that all components of the graph G_1 are isolated nodes. Thus $n+2 = 2k$ and there is not an edge in G_0 between any two nodes in G_1. □

We prove now some theorems about extended graphs of a general simplex.

Theorem 3.4.19 *Every signed graph, the node connectivity of whose positive part is at least 2 and which is transitive (i.e., for every two of its nodes u, v there exists an automorphism of the graph which transforms u into v), is an extended graph of some simplex.*

Proof. Denote by $0, 1, \ldots, n+1$ the nodes of the given graph G_0 and define two matrices $A = [a_{rs}]$ and $B = [b_{rs}]$, $r, s = 0, 1, \ldots, n+1$, of order $n+2$ as follows:

$a_{rs} = 1$, if $r \neq s$ and if r, s is a positive edge in G_0, otherwise, $a_{rs} = 0$;
$b_{rs} = 1$, if $r \neq s$ and if r, s is a negative edge in G_0, otherwise, $b_{rs} = 0$.

Denote by A_0 and B_0 the principal submatrices of A and B obtained by removing the row and column with index 0. The matrix A_0 is nonnegative and irreducible since its graph is connected. By the Perron–Frobenius theorem (cf. Theorem A.3.1), there exists a positive simple eigenvalue α_0 of A_0, which has of all eigenvalues the maximum modulus, and the corresponding eigenvector z_0 can be chosen positive

$$A_0 z_0 = \alpha_0 z_0.$$

It follows that the matrix $\alpha_0 I_0 - A_0$ (I_0 identity matrix) is positive semidefinite of rank n (and order $n+1$), and all its principal minors of orders $\le n$ are positive. This implies that there exists a number $\varepsilon > 0$ such that also the matrix

$$C_0 = A_0 - \varepsilon B_0$$

has a positive simple eigenvalue γ_0, for which there exists a positive eigenvector z

$$C_0 z = \gamma_0 z,$$

whereas the matrix

$$P_0 = \gamma_0 I_0 - C_0$$

is positive semidefinite of rank n. Observe that $P_0 z = 0$.

Form now the matrix

$$P = \gamma_0 I - A + \varepsilon B,$$

where I is the identity matrix of order $n+2$.

Using the transitivity of the graph G_0, we obtain that all principal minors of order $n+1$ of the matrix P are equal to zero, whereas all principal minors of orders $\le n$ are positive: indeed, if we remove from P the row and column with index 0, we obtain P_0 which has this property. If we remove from P a row and column with index $k > 0$, we obtain some matrix P_k; however, since there exists an automorphism of G_0 transforming the vertex 0 into the vertex k, there exists a permutation of rows and (simultaneously) columns of the matrix P_k, transforming it into P_0. Thus also $\det P_k = 0$ and all principal minors of P_k of degree $\le n$ are positive as well. We can show that for sufficiently small $\varepsilon > 0$, the matrix P is nonsingular: P cannot be positive definite (its principal minors of order $n+1$ are equal to zero) so that $\det P < 0$ and the signature of P is n. By Theorem 3.4.8, the negative of the signed graph of the matrix P is an extended graph of some n-simplex. According to the definitions of the matrices A and B, this is the graph G_0. □

Other important examples of extended graphs of n-simplexes are those of right simplexes. These simplexes will be studied independently in the next chapter.

Theorem 3.4.20 *Suppose a signed graph G_0 on $n+1$ nodes has the following properties:*

(i) *if we remove from G_0 one of its nodes u and the incident edges, the resulting graph is a tree T;*

(ii) *the node u is joined with each node v in T by a positive, or negative edge, according to whether v has in T degree 1, or ≥ 3 (and thus u is not joined to v if v has degree 2).*

Then G_0 is the extended graph of some n-simplex.

Proof. This follows immediately from Theorem 4.1.2 in the next chapter since G_0 is the extended graph of the right n-simplex having the usual graph T. □

Theorem 3.4.21 *Suppose G_0 is an extended graph of some n-simplex such that at least one of its nodes is saturated, i.e. it is joined to every other node by an edge (positive or negative). Then every signed supergraph of the graph G_0 (with the same set of nodes) is an extended graph of some n-simplex.*

Proof. Suppose Σ is an n-simplex, the extended graph of which is G_0; let the saturated node of G_0 correspond to the circumcenter C of Σ. Thus C is not contained in any face of Σ. If G_1 is any supergraph of G_0, G_1 has the same edges at the saturated node as G_0.

Let $[q_{rs}]$, $r, s = 0, \ldots, n+1$, be the matrix corresponding to Σ. The submatrix $Q = [q_{ij}]$, $i, j = 1, \ldots, n+1$, satisfies:

(i) Q is positive semidefinite of rank n;

(ii) $Qe = 0$, where $e = (1, \ldots, 1)^T$;

(iii) if $0, 1, \ldots, n+1$ are the nodes of the graph G_0 and 0 the saturated node, then for $i \neq j$, $i, j = 1, \ldots, n+1$,

$q_{ij} < 0$, if and only if (i, j) is positive in G_0,
$q_{ij} > 0$, if and only if (i, j) is negative in G_0.

Construct a new matrix $\tilde{Q} = [\tilde{q}_{ij}]$, $i, j = 1, \ldots, n+1$, as follows:
$\tilde{q}_{ij} = q_{ij}$, if $i \neq j$ and $q_{ij} \neq 0$;
$\tilde{q}_{ij} = -\varepsilon$, if $i \neq j$, $q_{ij} = 0$ and (i, j) is a positive edge of G_1;
$\tilde{q}_{ij} = \varepsilon$, if $i \neq j$, $q_{ij} = 0$ and (i, j) is a negative edge in G_1;
$\tilde{q}_{ii} = -\sum_{j \neq i} \tilde{q}_{ij}$.

We now choose the number ε positive and so small that $\tilde{Q}$ remains positive semidefinite and, in addition, such that the signs of the new numbers $\tilde{c}_i$ from (3.4.9) for the numbers $\tilde{q}_{ij}$ coincide with the signs of the numbers c_i from (3.4.9) for the numbers q_{ij}. Such a number $\varepsilon > 0$ clearly exists since all numbers c_i as barycentric coordinates of the circumcenter of Σ are different from zero.

It now follows easily that $\tilde{Q}$ is the Gramian of some n-simplex $\tilde{\Sigma}$, which has $\tilde{G}_0$ as its extended graph. □

Remark 3.4.22 This theorem can also be formulated as follows: Suppose G_0 is a signed graph on $n + 1$ nodes, which is not an extended graph of any n-simplex. Then no subgraph of G_0 with the same set of nodes and a saturated node can be the extended graph of an n-simplex.

We shall return to the topic of extended graphs in Chapter 4 in various classes of special simplexes and in Chapter 6, Section 4, where all possible extended graphs of tetrahedrons will be found.

4

Special simplexes

In this chapter, we shall study classes of simplexes with special properties in more detail.

4.1 Right simplexes

We start with the following lemma.

Lemma 4.1.1 *Let the graph $G = (V, E)$ be a tree with the node set $V = \{1, 2, \dots, n+1\}$ and edge set E. Assign to every edge (i, k) in E a nonzero number c_{ik}. Denote by Γ the matrix $[\gamma_{rs}]$, $r, s = 0, 1, \dots, n+1$, with entries $\gamma_{00} = \sum_{ik \in G} 1/c_{ik}$, $\gamma_{0i} = \gamma_{i0} = s_i - 2$, where s_i is the degree of the node i in G (i.e. the number of edges incident with i)*

$$\begin{aligned} \gamma_{ik} = \gamma_{ki} &= -c_{ik}, \quad \text{if } i \neq k, (i,k) \in E, \\ \gamma_{ik} &= 0 \quad \text{for } i \neq k, (i,k) \notin E, \\ \gamma_{ii} &= \sum_{k, (i,k) \in E} c_{ik}. \end{aligned}$$

Further, denote by M the matrix $[m_{rs}]$, $r, s = 0, \dots, n+1$, with entries $m_{00} = 0$, $m_{0i} = m_{i0} = 1$, $m_{ii} = 0$

$$m_{ik} = m_{ki} = \frac{1}{c_{ij_1}} + \frac{1}{c_{j_1 j_2}} + \dots + \frac{1}{c_{j_s k}},$$

if $i \neq k$ and $(i, j_1, \dots, j_s, k)$ is the (unique) path in G from the node i to k.
Then

$$M\Gamma = -2I,$$

where I is the identity matrix of order $n+2$. (In the formulae above, we write again i, j, k for indices $1, 2, \dots, n+1$.)

Proof. We have to show that

$$\sum_{s=0}^{n+1} m_{rs}\gamma_{st} = -2\delta_{rt}.$$

We begin with the case $r = t = 0$. Then
$\sum_{s=0}^{n+1} m_{0s}\gamma_{s0} = \sum_{i=1}^{n+1} m_{0i}\gamma_{0i} = \sum_{i=1}^{n+1}(s_i - 2) = -2$, since $\sum_{i=1}^{n+1} s_i = 2n$ ($\sum s_i$ is twice the number of edges of G and the number of edges of a tree is by one less than the number of nodes).

For $r = 0$, $t = i = 1, \ldots, n$ we have

$$\sum_{s=0}^{n+1} m_{0s}\gamma_{si}\Big(= \sum_{k=1}^{n+1}\gamma_{ik}\Big) = 0,$$

whereas for $r = i$, $t = 0$

$$\sum_{s=0}^{n+1} m_{is}\gamma_{0s} = \sum_{jl\in E}\frac{1}{c_{jl}} + \sum_{k\neq i}(s_k - 2)\left(\frac{1}{c_{ij_1}} + \frac{1}{c_{j_1j_2}} + \cdots + \frac{1}{c_{j_sk}}\right).$$

To show that the sum on the right-hand side is zero, let us prove that in the sum

$$\sum_{k\neq i}(s_k - 2)\left(\frac{1}{c_{ij_1}} + \frac{1}{c_{j_1j_2}} + \cdots + \frac{1}{c_{j_sk}}\right)$$

the term $1/c_{jl}$ appears for every edge (j, l) with the coefficient -1. Thus let $(j, l) \in E$; the node i is in one of the parts V_j, V_l obtained by deleting the edge (j, l) from E. Suppose that i is in V_j, say. Then $1/c_{jl}$ appears in those summands k, which belong to the branch V_l containing l, namely with the total coefficient

$$\sum_{k\in V_l}(s_k - 2).$$

Let p be the number of nodes in V_l. Then $\sum_{k\in V_l} s_k = 1 + 2(p-1)$ (the node l has degree s_l by one greater than that in V_l, and the sum of the degrees of all the nodes is twice the number of edges). Consequently, $\sum_{k\in V_l}(s_k - 2) = 2(p-1) - 2p + 1 = -1$.

For $r = i$, $t = i$, we obtain

$$\begin{aligned}\sum_{s=0}^{n+1} m_{is}\gamma_{si} &= \gamma_{0i} + \sum_{j, j\neq i} m_{ij}\gamma_{ij}\\ &= s_i - 2 - \sum_{(i,j)\in E}\frac{1}{c_{ij}}c_{ij}\\ &= s_i - 2 - s_i\\ &= -2.\end{aligned}$$

Finally, if $r = i$, $t = j \neq i$, we have

$$\begin{aligned}\sum_{s=0}^{n+1} m_{is}\gamma_{sj} &= m_{0i}\gamma_{0j} + m_{ij}\gamma_{jj} + \sum_{k,i\neq k\neq j} m_{ik}\gamma_{kj}\\ &= s_j - 2 + m_{ij}\sum_{l,(j,l)\in E} c_{jl} - m_{ij}\sum_{k,(j,k)\in E} c_{kj} + m_{jt}c_{tj}\\ &\quad - \sum_{l\neq t,(j,l)\in E} m_{jl}c_{jl},\end{aligned}$$

where t is the node neighboring j in the path from i to j, since $m_{it} = m_{ij} - m_{jt}$, whereas for the remaining neighbors l of j, $m_{il} = m_{ij} + m_{jl}$.

This confirms that the last expression is indeed equal to zero. □

In Definition 3.1.6, we introduced the notion of a right n-simplex as such an n-simplex, which has exactly n acute interior angles and all of the remaining interior angles ($\binom{n}{2}$ in number) right.

The signed graph G_Σ (cf. Corollary 3.1.7) of such a right n-simplex is thus a tree with all edges positive. In the colored form (cf. Definition 3.1.8), these edges are colored red. In agreement with usual notions for the right triangle, we call *cathetes*, or *legs*, the edges opposite to acute interior angles, whereas the *hypotenuse* of the simplex will be the face containing exactly those vertices which are incident with one leg only.

We are now able to prove the main theorem on right simplexes ([6]). It also gives a hint for an easy construction of such a simplex.

Theorem 4.1.2 (Basic theorem on right simplexes)

(i) *Any two legs of a right n-simplex are perpendicular to each other; they form thus a cathetic tree.*

(ii) *The set of $n+1$ vertices of a right n-simplex can be completed to the set of 2^n vertices of a rectangular parallelepiped (we call it simply a right n-*box*) in $\mathcal{E}_n$, namely in such a way that the legs are (mutually perpendicular) edges of the box.*

(iii) *Conversely, if we choose among the $n2^{n-1}$ edges of a right n-box a connected system of n mutually perpendicular edges, then the vertices of the edges (there are $n+1$ of them) form a right n-simplex whose legs coincide with the chosen edges.*

(iv) *The barycentric coordinates of the center of the circumscribed hypersphere of a right n-simplex are $1 - \frac{1}{2}s_i$, where s_i is the degree of the vertex A_i in the cathetic tree.*

Proof. Let a right n-simplex Σ be given. The numbers q_{ik}, i.e. the entries of the Gramian of this simplex, satisfy all assumptions for the numbers γ_{ik} in

Theorem 4.1.1; the graph $G = (V, E)$ from Theorem 4.1.1 is the graph of the simplex.

The angles φ_{ik} defined by

$$\cos\varphi_{ik} = -\frac{q_{ik}}{\sqrt{q_{ii}}\sqrt{q_{kk}}},$$

are then the interior angles of the simplex Σ.

Let m_{rs} be the entries of the matrix M from Theorem 4.1.1. We intend to show that for $i, k = 1, \ldots, n+1$, the numbers m_{ik} are squares of the lengths of edges of the simplex Σ.

To this end, we construct in some Euclidean n-space $\mathcal{E}_n$ with the usual orthonormal basis $e_1, \ldots, e_n$ an n-simplex $\tilde{\Sigma}$.

Observe first that the numbers q_{ik}, for $(i, k) \in E$, are always negative since $q_{ik} = -p_i p_k \cos\varphi_{ik}$, where $p_i > 0$ and $\cos\varphi_{ik} > 0$.

Choose now a point A_{n+1} arbitrarily in $\mathcal{E}_n$ and define the points A_i, $i = 1, \ldots, n$ in $\mathcal{E}_n$ as follows:

Number the n edges of the graph G in some way by numbers $1, 2, \ldots, n$. If $(n+1, j_1, \ldots, j_s, i)$ is the path from the node $n+1$ to the node i in the graph G, set

$$A_i = A_{n+1} + \frac{1}{\sqrt{-q_{n+1,j_1}}} e_{c_1} + \frac{1}{\sqrt{-q_{j_1,j_2}}} e_{c_2} + \cdots + \frac{1}{\sqrt{-q_{j_k,i}}} e_{c_k},$$

where c_1 is the number assigned to the edge $(n+1, j_1)$, c_2 the number assigned to the edge (j_1, j_2), etc.

The squares $\widehat{m}_{ik}$ of the distances of the points A_i and A_k then satisfy

$$\widehat{m}_{ik} = -\frac{1}{q_{ik}} \qquad \text{if } (i, k) \in G,$$

and, since there is a unique path between nodes of G, we have in general

$$\widehat{m}_{ik} = -\Big(\frac{1}{q_{ij_1}} + \frac{1}{q_{j_1 j_2}} + \cdots + \frac{1}{q_{j_s k}}\Big),$$

if $(i, j_1, \ldots, j_s, k)$ is the path from i to k in G.

The points $A_1, \ldots, A_{n+1}$ are linearly independent in $\mathcal{E}_n$, thus forming vertices of an n-simplex $\widehat{\Sigma}$.

The corresponding numbers $\tilde{q_{ik}}$ and $\tilde{q_{0i}}$ of the Gramian of this n-simplex are clearly equal to the numbers γ_{ik} and γ_{0i}. By Theorem 2.1.1, the interior angles of $\widehat{\Sigma}$ and Σ are the same and these simplexes are similar.

The vertices $A_1, \ldots, A_{n+1}$ of the n-simplex $\widehat{\Sigma}$ are indeed contained in the set of 2^n vertices of the right n-box

$$P(\varepsilon_1, \ldots, \varepsilon_n) = A_{n+1} + \sum_{i=1}^{n} \varepsilon_i \frac{e_i}{\sqrt{-q_{pq}}},$$

where $\varepsilon_i = 0$ or 1 and $-q_{pq}$ is the weight of that edge of the graph G, which was numbered by i.

It follows that the same holds for the given simplex Σ. It also follows that the legs of Σ are perpendicular.

To prove the last assertion (iv), observe that by Theorem 2.1.1, the numbers $\tilde{q_{0i}}$ are homogeneous barycentric coordinates of the circumcenter of $\widehat{\Sigma}$, and these are proportional to the numbers $1 - \frac{1}{2}s_i$, the sum of which is already 1. □

Observe that a right n-simplex is (uniquely up to congruence) determined by its cathetic tree, i.e. the structure and the lengths of the legs. There is also an intimate relationship between right simplexes and weighted trees, as the following theorem shows.

Theorem 4.1.3 *Let G be a tree with $n+1$ nodes $U_1, \ldots, U_{n+1}$. Let each edge $(U_p, U_q) \in G$ be assigned a positive number $\mu(U_p, U_q)$; we call it the length of the edge. Define the distance $\mu(U_i, U_k)$ between an arbitrary pair of nodes U_i, U_k as the sum of the lengths of the edges in the path between U_i and U_k.*

Then there exists an n-simplex with the property that the squares m_{ik} of the lengths of edges satisfy

$$m_{ik} = \mu(U_i, U_k). \tag{4.1}$$

This n-simplex is a right n-simplex and its cathetic tree is isomorphic to G.

Conversely, for every right n-simplex there exists a graph G isomorphic to the cathetic tree of the simplex and a metric μ on G such that (4.1) *holds.*

Proof. First let a tree G with the prescribed properties be given. Then there exists in some Euclidean n-space $\mathcal{E}_n$ a right box and a tree T isomorphic to G consisting of edges, no two of them being parallel and such that the lengths of the edges of the box are equal to the square roots of the corresponding numbers $\mu(U_i, U_k)$. The nodes $A_1, \ldots, A_{n+1}$ of T then satisfy

$$\rho^2(A_i, A_k) = \mu(U_i, U_k)$$

for all $i, k = 1, \ldots, n+1$; they are not in a hyperplane, and thus form vertices of a right n-simplex with the required properties.

Conversely, let Σ in $\mathcal{E}_n$ be a right n-simplex. Then there exists in $\mathcal{E}_n$ a right box, n edges of which coincide with the cathetes of Σ. By the Pythagorean theorem, the tree isomorphic to the cathetic tree, each edge of which is assigned the square of the length of the corresponding cathete, has the property (4.1). □

An immediate corollary of Theorem 4.1.2 is the following:

Theorem 4.1.4 *The volume of a right n-simplex is the $(1/n!)$-multiple of the product of the lengths of the legs.*

Theorem 4.1.5 *The hypotenuse of a right simplex is a strictly acute face, which among strictly acute faces has the maximum dimension.*

Proof. Suppose Σ is a right n-simplex so that the graph G_Σ is a tree. Let U be the set of nodes of G_Σ, and U_1 be the subset of U consisting of all nodes of degree one of G_Σ. Let $m = |U_1|$. It follows immediately from the property of the elimination graph obtained by eliminating the nodes from $U \setminus U_1$ that the graph of the hypotenuse is complete.

To prove the maximality, suppose there is in Σ a strictly acute face F with more than m vertices. Denote by H the hypotenuse.

Case 1. F contains H. Let u be a vertex in F which is not in H. Since removing u from G_Σ leads to at least two components, there exist two vertices p, q in H, which are in such different components. The triangle with vertices p, q, and u has a right angle at u, a contradiction.

Case 2. There is a vertex v in H which is not in F. The face F_0 of Σ opposite to v is a right $(n-1)$-simplex Σ'. Its hypotenuse has at most m vertices and is a strictly acute face of Σ' having the maximal dimension. This is a contradiction to the fact that F is contained in Σ'. □

In the last theorem of this section, we need the following lemma whose geometric interpretation is left to the reader.

Lemma 4.1.6 *Suppose that in a hyperacute n-simplex with vertices A_i and $(n-1)$-dimensional faces ω_i, the angle $A_iA_jA_k$ is right. Then the node u_j of the graph G of the simplex (corresponding to the face ω_j) is a cut-node in G, which separates the nodes u_i and u_k (corresponding to ω_i and ω_k).*

Proof. The fact that $A_iA_jA_k$ is a right triangle with the right angle at A_j means that there is no path in G from u_i to u_k not containing u_j. Therefore, u_i and u_k are in different components of the graph G' obtained from G by deleting the node u_j and the outgoing edges. Thus G' is not connected. □

We now summarize properties of one special type of right simplex.

Theorem 4.1.7 *Let Σ be an n-simplex. Then the following are equivalent:*

(i) *The signed graph of Σ is a positive path.*
(ii) *There exist positive numbers $a_1, \ldots, a_n$ and a Cartesian coordinate system such that the vertices of Σ can be permuted into the position as in Example 3.1.10.*
(iii) *There exist distinct real numbers $c_1, \ldots, c_{n+1}$ such that the squares of the lengths of edges m_{ik} satisfy*

$$m_{ik} = |c_i - c_k|. \tag{4.2}$$

(iv) *All two-dimensional faces of Σ are right triangles.*

(v) *Σ is a hyperacute n-simplex with the property that its circumcenter is a point of its edge.*

(vi) *The Gramian of Σ is a permutation of a tridiagonal matrix.*

Proof. (i) $\leftrightarrow$ (ii). Suppose (i). Then Σ is a right simplex and the construction from Theorem 4.1.2 yields immediately (ii). Conversely, we saw in Example 3.1.10 that (ii) implies (i).

(ii) $\leftrightarrow$ (iii). Given (ii), define $c_1 = 0$, $c_i = \sum_{k=1}^{i-1} a_k^2$. Then (iii) will be satisfied. Conversely, if the vertices of Σ in (iii) are renumbered so that $c_1 < c_2 < \cdots < c_{n+1}$, then $a_i = \sqrt{c_{i+1} - c_i}$ is realized as the simplex in Example 3.1.10.

(iii) $\leftrightarrow$ (iv). Suppose (iii). If A_i, A_j, A_k are distinct vertices, the distances satisfy the Pythagorean equality. Let us prove that (iv) implies (iii). Suppose that all two-dimensional faces of an n-simplex Σ with vertices $A_1, \ldots, A_{n+1}$ are right triangles. Observe first that if A_iA_j is one from the set of edges of Σ with maximum length, then it is the hypotenuse in all triangles $A_iA_jA_k$ for $i \neq k \neq j$. This means that the hypersphere having A_iA_j as the diameter is the circumscribed sphere of the simplex. Therefore, just one such maximum edge exists.

Let now A_i, A_j, A_k, A_l be four distinct vertices and let A_iA_j have the maximum length of the six edges connecting them. We show that the angle $\angle A_kA_iA_l$ cannot be right. Suppose to the contrary that $\angle A_kA_iA_l = \frac{1}{2}\pi$.

Distinguish two cases:

Case A. $\angle A_kA_jA_l = \frac{1}{2}\pi$; this implies that the midpoint of the edge A_kA_l has the same distance from all four vertices considered, a contradiction to the above since the midpoint of A_iA_j has this property.

Case B. One of the angles $\angle A_kA_lA_j$, $\angle A_lA_kA_j$ is right. Since both these cases differ by changing the indices k and l only, we can suppose that $\angle A_kA_lA_j = \frac{1}{2}\pi$. We have then by the Pythagorean theorem that

$$\begin{aligned} m_{ik} + m_{jk} &= m_{ij}, \quad & m_{il} + m_{jl} &= m_{ij}, \\ m_{ik} + m_{il} &= m_{kl}, \quad & m_{kl} + m_{jl} &= m_{jk}. \end{aligned}$$

These imply that

$$m_{ik} + m_{il} + m_{jl} = m_{kl} + m_{jl} = m_{jk},$$

as well as

$$m_{ik} + m_{il} + m_{jl} = m_{ik} + m_{ij} = 2m_{ik} + m_{jk}, \quad \text{i.e. } m_{ik} = 0.$$

This contradiction shows that $\angle A_kA_iA_l \neq \frac{1}{2}\pi$.

Suppose that A_iA_j is the edge of the simplex of maximum length. Choose an arbitrary real number c_i and set for each $k = 1, 2, \ldots, n+1$

$$c_k = c_i + m_{ik}. \tag{4.3}$$

Let us prove that (4.2) holds and that all the numbers c_k are distinct. Indeed, if $c_k = c_l$ for $k \neq l$, then necessarily $j \neq k, j \neq l$ (A_iA_j is the only longest edge) and, as was shown above, $\angle A_kA_iA_l \neq \frac{1}{2}\pi$, i.e. exactly one of the edges A_iA_k, A_iA_l is the hypotenuse in the right triangle $A_iA_kA_l$, i.e. $m_{il} \neq m_{jl}$, contradicting $c_k = c_l$. By (4.3)

$$m_{ik} = c_k - c_i = |c_k - c_i|; \tag{4.4}$$

since $m_{ij} = m_{ik} + m_{jk}$ for all k, it follows that also

$$m_{jk} = c_j - c_k = |c_j - c_k|. \tag{4.5}$$

Suppose now that k, l, i, j are distinct indices. Then $\angle A_kA_iA_l \neq \frac{1}{2}\pi$, so that either $m_{ik} + m_{kl} = m_{il}$ or $m_{il} + m_{kl} = m_{ik}$. In the first case, $m_{kl} = c_l - c_i - c_k + c_i = c_l - c_k$; in the second, $m_{kl} = c_k - c_l$. Thus in both cases

$$m_{kl} = |c_k - c_l|. \tag{4.6}$$

The equations (4.4), (4.5), and (4.6) imply (4.2).

(ii)→ (v). This was shown in Example 3.1.10.

(v)→ (i). Suppose now that $A_1, \ldots, A_{n+1}$ are the vertices of a hyperacute simplex Σ and $\omega_1, \ldots, \omega_{n+1}$ are its $(n-1)$-dimensional faces, ω_i opposite to A_i. Let the circumcenter be a point of the edge A_1A_{n+1}, say. By the Thalet theorem, all the angles $\angle A_1A_jA_{n+1}$ are right for all j, $1 < j < n+1$. Lemma 4.1.6 implies that every vertex u_j corresponding to ω_j, $1 < j < n+1$, is a cut-vertex of the graph G_Σ separating the vertices u_1 and u_{n+1}.

We intend to show that G_Σ is a path between u_1 and u_{n+1}. Since G_Σ is connected, there exists a path P in G_Σ from u_1 to u_{n+1}, with the minimum number of vertices. Suppose that a vertex u_k (corresponding to ω_k) is not in P. Since u_k separates u_1 and u_{n+1} in G_Σ, it is a contradiction. Minimality of P then implies that G_Σ is P and Σ satisfies (i).

(i) ↔ (vi). This follows from Corollary 3.2.1 and the notion of the tridiagonal matrix (cf. Appendix). □

We call a simplex that satisfies any one of the conditions (i)–(vi) a *Schlaefli simplex* since Schlaefli used this simplex (he called it *Orthoscheme*) in his study of volumes in noneuclidean spaces.

We now have the last result of this section.

Theorem 4.1.8 *Every face of dimension at least two of a Schlaefli simplex is again a Schlaefli simplex.*

Proof. This follows immediately from (ii) of Theorem 4.1.7. □

4.2 Orthocentric simplexes

Whereas altitudes of a triangle meet in one point, this is no longer true in general for the tetrahedron. In this section, we characterize those simplexes for which the altitudes meet in one point; they are called *orthocentric* and they were studied before (cf. [2]). We suppose that the dimension is at least 2.

Theorem 4.2.1 *Let Σ be an orthocentric n-simplex. Then the squares m_{ik} of the lengths of edges have the property that there exist real numbers $\pi_1, \dots, \pi_{n+1}$ such that for all i, k, $i \neq k$*

$$m_{ik} = \pi_i + \pi_k. \tag{4.7}$$

In addition, the numbers π_i satisfy the following:

(i) *either all of them are positive (and the simplex is* acute orthocentric*), or*
(ii) *one of the numbers π_i is zero and the remaining positive (the simplex is* right orthocentric*), or finally*
(iii) *one of the numbers π_i is negative, and the remaining positive (the simplex is* obtuse orthocentric*), and*

$$\sum_{k=1}^{n+1} \frac{1}{\pi_k} < 0. \tag{4.8}$$

Conversely, if $\pi_1, \dots, \pi_{n+1}$ are real numbers for which one of the conditions (i), (ii), (iii) *holds, then there exists an n-simplex, whose squares of lengths of edges satisfy* (4.7)*, and this simplex is orthocentric. The intersection point V of the altitudes, i.e. the* orthocenter*, has homogeneous barycentric coordinates $V \equiv (v_i)$, $v_i = \prod_{k=1, k\neq i}^{n+1} \pi_k$; i.e., in cases* (i) *and* (iii)*, $V \equiv (1/\pi_i)$.*

Proof. Let Σ with vertices $A_1, \dots, A_{n+1}$ have orthocenter V. Denote by c_i the vectors

$$c_i = A_i - A_{n+1}, \qquad i = 1, \dots, n, \tag{4.9}$$

by d_i the vectors

$$d_i = A_i - V, \qquad i = 1, \dots, n+1. \tag{4.10}$$

The vector d_i, $i \in \{1, \dots, n+1\}$, is either the zero vector, or it is perpendicular to the face ω_i, and therefore to all vectors in this face.

We have thus

$$\langle d_i, c_k \rangle = 0 \quad \text{for} \quad k \neq i,\ i, k = 1, \dots, n. \tag{4.11}$$

The vector d_{n+1} is also either the zero vector, or it is perpendicular to all vectors $\overrightarrow{A_i A_k} = c_k - c_i$ for $i \neq k$, $i, k = 1, \dots, n$. Thus, in all cases

$$\langle d_{n+1}, c_k - c_i \rangle = 0, \qquad i \neq k,\ i, k = 1, \dots, n.$$

Denote now $\langle d_{n+1}, c_1\rangle = -\pi_{n+1}$; we obtain

$$\langle d_{n+1}, c_k\rangle = -\pi_{n+1}, \qquad k = 1, \ldots, n.$$

Denote also

$$\langle d_k, c_k\rangle = \pi_k, \qquad k = 1, \ldots, n.$$

We intend to show that (4.7) holds.

First, observe that by (4.9) and (4.10)

$$c_i - c_j = d_i - d_j, \qquad i, j = 1, \ldots, n,$$

$$c_i = d_i - d_{n+1}, \qquad i = 1, \ldots, n.$$

Therefore, we have for $i = 1, \ldots, n$, that

$$\begin{aligned} m_{i,n+1} &= \langle c_i, c_i\rangle \\ &= \langle c_i, d_i - d_{n+1}\rangle \\ &= \langle c_i, d_i\rangle - \langle c_i, d_{n+1}\rangle \\ &= \pi_i + \pi_{n+1}. \end{aligned}$$

If $i \neq k$, $i, k = 1, \ldots, n$, then (4.11) implies that

$$\begin{aligned} m_{ik} &= \langle c_i - c_k, c_i - c_k\rangle \\ &= \langle c_i - c_k, d_i - d_k\rangle \\ &= \langle c_i, d_i\rangle + \langle c_k, d_k\rangle \\ &= \pi_i + \pi_k. \end{aligned}$$

The relations (4.7) are thus established. Since $\pi_i + \pi_k > 0$ for all $i, k = 1, \ldots, n+1$, $i \neq k$, at most one of the numbers π_i is not positive.

Before proving the condition (4.8) in case (iii), express the quadratic form $\sum m_{ik}\, x_i\, x_k$ as

$$\begin{aligned} \sum_{i,k=1}^{n+1} m_{ik}\, x_i\, x_k &= \sum_{i,k=1, i\neq k} (\pi_i + \pi_k) x_i\, x_k \\ &= \sum_{i,k=1}^{n+1} (\pi_i + \pi_k) x_i\, x_k - 2 \sum_{i=1}^{n+1} \pi_i\, x_i^2, \end{aligned}$$

or

$$\sum_{i,k=1}^{n+1} m_{ik}\, x_i\, x_k = -2 \sum_{i=1}^{n+1} \pi_i\, x_i^2 + 2 \sum_{i=1}^{n+1} \pi_i\, x_i \sum_{i=1}^{n+1} x_i. \tag{4.12}$$

Suppose now that one of the numbers π_i, say π_{n+1}, is negative. Then $\pi_i > 0$ for $i = 1, \ldots, n$. By Theorem 1.2.4, $\sum m_{ik}\, x_i\, x_k < 0$ for $x_i = 1/\pi_i$, $i = 1, \ldots, n$, $x_{n+1} = -\sum_{i=1}^{n} (1/\pi_i)$. By (4.12), we obtain

$$\sum_{i,k=1}^{n+1} m_{ik}\, x_i\, x_k = -2\Big(\sum_{i=1}^{n} \pi_i \frac{1}{\pi_i^2} + \pi_{n+1}\Big(\sum_{i=1}^{n} \frac{1}{\pi_i}\Big)^2\Big)$$
$$= -2\sum_{i=1}^{n} \frac{1}{\pi_i}\Big(1 + \pi_{n+1}\sum_{i=1}^{n} \frac{1}{\pi_i}\Big)$$
$$= 2|\pi_{n+1}|\sum_{i=1}^{n} \frac{1}{\pi_i}\Big(\sum_{k=1}^{n+1} \frac{1}{\pi_i}\Big).$$

This proves (4.8).

Suppose now that $\pi_1, \dots, \pi_{n+1}$ are real numbers which fulfill one of the conditions (i), (ii), or (iii). If (iii) or (ii) holds, then, by (4.12), whenever $\sum_{i=1}^{n+1} x_i = 0$ and $m_{ik} = \pi_i + \pi_k$, $i \neq k$, $i, k = 1, \dots, n+1$

$$\sum_{i,k=1}^{n+1} m_{ik}\, x_i\, x_k = -2\sum_{i=1}^{n+1} \pi_i\, x_i^2 \leq 0.$$

Equality is attained only if $x_1 = \cdots = x_{n+1} = 0$. By Theorem 1.2.4, the n-simplex really exists.

Suppose now that condition (iii) holds, i.e. $x_{n+1} < 0$, together with the inequality (4.8). Suppose that x_i are real numbers not all equal to zero and such that $\sum_{i=1}^{n+1} x_i = 0$. The numbers $m_{ik} = \pi_i + \pi_k$, $i \neq k$, $i, k = 1, \dots, n+1$ satisfy

$$\sum_{i,k=1}^{n+1} m_{ik}\, x_i\, x_k = -2\sum_{i=1}^{n+1} \pi_i\, x_i^2 = -2\Big[\sum_{i=1}^{n} \pi_i x_i^2 + \pi_{n+1}\Big(\sum_{i=1}^{n} x_i\Big)^2\Big].$$

By Schwarz inequality

$$\sum_{i=1}^{n} \frac{1}{\pi_i} \sum_{i=1}^{n} \pi_i\, x_i^2 \geq \Big(\sum_{i=1}^{n} \frac{1}{\sqrt{\pi_i}}\sqrt{\pi_i} x_i\Big)^2 = \Big(\sum_{i=1}^{n} x_i\Big)^2.$$

Therefore, since $\pi_{n+1} < 0$

$$\sum_{i,k=1}^{n+1} m_{ik}\, x_i\, x_k \leq -2\Big[\sum_{i=1}^{n} \pi_i x_i^2 + \pi_{n+1}\sum_{i=1}^{n} \frac{1}{\pi_i} \sum_{i=1}^{n} \pi_i\, x_i^2\Big]$$
$$= 2|\pi_{n+1}|\sum_{i=1}^{n} \pi_i\, x_i^2\Big(\frac{1}{\pi_{n+1}} + \sum_{i=1}^{n} \frac{1}{\pi_i}\Big) < 0.$$

By Theorem 1.2.4 again, there exists an n-simplex satisfying $m_{ik} = \pi_i + \pi_k$ for $i \neq k$. It remains to show that in all cases (i), (ii), and (iii), the point $V \equiv (v_i)$, $v_i = \prod_{k \neq i} \pi_k$, is the orthocenter. If $\pi_{n+1} = 0$, $V \equiv A_{n+1}$ and

$m_{ik} = m_{i,n+1} + m_{k,n+1}$ for $i \neq k$, $i,k = 1,\ldots,n$. The vectors $\overrightarrow{A_iA_{n+1}}$ are thus mutually perpendicular and coincide with the altitudes orthogonal to the faces ω_i, $i = 1,\ldots,n$. Since also the last altitude contains the point A_{n+1}, $A_{n+1} \equiv V$ is indeed the orthocenter.

Suppose now that all π_is are different from zero. Let us show that the vector $\overrightarrow{VA_i}$, where $V \equiv (1/\pi_i)$, is perpendicular to all vectors $\overrightarrow{A_jA_k}$ for $i \neq j \neq k \neq i$, $i,j,k = 1,\ldots,n+1$. As we know from (1.10), the vectors p, q are perpendicular if and only if $\sum m_{ik}p_iq_k = 0$. By (4.12), since for such vectors $\sum p_i = \sum q_i = 0$

$$\sum m_{ik}p_iq_k = -2\sum_{i=1}^{n+1} \pi_i\, p_i\, q_i. \tag{4.13}$$

Further, if we simply denote $\sum\limits_{k=1}^{n+1}(1/\pi_i)$ by τ

$$\overrightarrow{VA_i} = (p_k), \qquad p_k = \tau\delta_{ik} - \frac{1}{\pi_k},$$

where δ_{ik} is the Kronecker delta.

Now

$$\overrightarrow{A_jA_k} = (q_l), \qquad q_l = \delta_{jl} - \delta_{kl},$$

so that, by (4.13)

$$\begin{aligned} -\frac{1}{2}\sum m_{ik}p_iq_k &= \sum_{s=1}^{n+1} \pi_s\Big(\tau\delta_{is} - \frac{1}{\pi_s}\Big)(\delta_{js} - \delta_{ks}) \\ &= \tau\sum_{s=1}^{n+1} \pi_s\,\delta_{is}(\delta_{js} - \delta_{ks}) - \sum_{s=1}^{n+1}(\delta_{js} - \delta_{ks}). \end{aligned}$$

Since both summands are zero, $\overrightarrow{VA_i} \perp \omega_i$, and V is the orthocenter. □

The second part could have been proved using the numbers q_{ik} in (1.21), which are determined by the numbers m_{ik} in formula (4.7). Let us show that if all the numbers π_i are different from zero, the numbers q_{rs} satisfying (1.21) are given by

$$\begin{aligned} \rho q_{00} &= \sum_{k=1}^{n+1} \pi_k \sum_{k=1}^{n+1} \frac{1}{\pi_k} - (n-1)^2, \\ \rho q_{0i} &= \frac{n-1}{\pi_i} - \sum_{k=1}^{n+1} \frac{1}{\pi_k}, \\ \rho q_{ii} &= \frac{1}{\pi_i}\Big(\sum_{k=1}^{n+1} \frac{1}{\pi_k} - \frac{1}{\pi_i}\Big), \\ \rho q_{ij} &= -\frac{1}{\pi_i\pi_j} \quad \text{for } i \neq j, \end{aligned} \tag{4.14}$$

where $\rho = \sum_{k=1}^{n+1}(1/\pi_k)$; observe that ρ is positive in case (i) and negative in case (iii) of Theorem 4.2.1.

Indeed

$$\begin{aligned}\rho\sum_{i=0}^{n+1} m_{0r}q_{i0} &= (n-1)\sum_{k=1}^{n+1}\frac{1}{\pi_k} - (n+1)\sum_{k=1}^{n+1}\frac{1}{\pi_k}\\ &= -2\sum_{k=1}^{n+1}\frac{1}{\pi_k}\\ &= -2\rho,\end{aligned}$$

so that $\sum_{r=0}^{n+1} m_{0r}\, q_{k0} = -2$.

Further, for $i, j, k, l = 1, \ldots, n+1$

$$\begin{aligned}\rho\sum_{r=0}^{n+1} m_{0r}q_{ri} &= \rho\Big(q_{ii} + \sum_{k\neq i} q_{ki}\Big)\\ &= \frac{1}{\pi_i}\left(\sum\frac{1}{\pi_k} - \frac{1}{\pi_i}\right) - \frac{1}{\pi_i}\sum_{k\neq i}\frac{1}{\pi_k}\\ &= 0,\end{aligned}$$

as well as

$$\begin{aligned}\rho\sum_{r=0}^{n+1} m_{ir}q_{r0} &= \rho\Big(q_{00} + \sum_{k\neq i} m_{ik}q_{k0}\Big)\\ &= \sum\pi_k\sum\frac{1}{\pi_k} - (n-1)^2\\ &\quad + \sum_{k\neq i}(\pi_i+\pi_k)\left(\frac{n-1}{\pi_k} - \sum\frac{1}{\pi_k}\right) = 0,\end{aligned}$$

since the last summand is equal to

$$\begin{aligned}(n-1)\pi_i\sum_{k\neq i}\frac{1}{\pi_k} &+ n(n-1) - n\pi_i\sum\frac{1}{\pi_l} - \sum_{k\neq i}\pi_k\sum\frac{1}{\pi_l}\\ &= (n-1)^2 - \sum\pi_k\sum\frac{1}{\pi_k}.\end{aligned}$$

Finally, for $i \neq k$

$$\begin{aligned}\rho\sum_{r=0}^{n+1} m_{ir}q_{rk} &= \rho\Big(q_{0k} + m_{ik}q_{kk} + \sum_{j\neq i, j\neq k} m_{ij}q_{jk}\Big)\\ &= \frac{n-1}{\pi_k} - \sum\frac{1}{\pi_l} + (\pi_i+\pi_k)\frac{1}{\pi_k}\sum_{l\neq k}\frac{1}{\pi_l}\\ &\quad - \sum_{j\neq i, j\neq k}(\pi_i+\pi_j)\frac{1}{\pi_j\pi_k}\end{aligned}$$

$$= \frac{n-1}{\pi_k} + \frac{\pi_i}{\pi_k}\sum_{l\neq k}\frac{1}{\pi_l} - \frac{1}{\pi_k} - \frac{\pi_i}{\pi_k}\sum_{j\neq i, j\neq k}\frac{1}{\pi_j} - (n-1)\frac{1}{\pi_k}$$

$$= \frac{\pi_i}{\pi_k}\frac{1}{\pi_i} - \frac{1}{\pi_k} = 0,$$

as well as

$$\begin{aligned}
\rho\sum_{r=0}^{n+1} m_{ir}q_{ri} &= \rho\left(q_{0i} + \sum_{k\neq i} m_{ik}q_{ki}\right)\\
&= \frac{n-1}{\pi_i} - \sum\frac{1}{\pi_l} - \sum_{k\neq i}(\pi_i+\pi_k)\frac{1}{\pi_i\pi_k}\\
&= \frac{n-1}{\pi_i} - \sum\frac{1}{\pi_l} - \sum_{k\neq i}\frac{1}{\pi_k} - \frac{n}{\pi_i}\\
&= -2\sum\frac{1}{\pi_l} = -2\rho,
\end{aligned}$$

as we wanted to prove.

The formulae (4.14) imply that in the case that all the π_is are nonzero, the point $V \equiv (1/\pi_i)$ belongs to the line joining the point A_k with the point $(q_{k1}, \dots, q_{k,n+1})$; however, this is the improper point, namely the direction perpendicular to the face ω_k. It follows that V is the orthocenter.

In the case that $\pi_{n+1} = 0$, we obtain, instead of formulae (4.14)

$$\begin{aligned}
q_{00} &= \sum_{k=1}^{n}\pi_k,\\
q_{0i} &= -1, \qquad i = 1, \dots, n,\\
q_{0,n+1} &= n-2,\\
q_{ii} &= \frac{1}{\pi_i}, \qquad i = 1, \dots, n,\\
q_{n+1,n+1} &= \sum_{k=1}^{n}\frac{1}{\pi_k},\\
q_{ij} &= 0, \qquad i \neq j,\ i, j = 1, \dots, n,\\
q_{i,n+1} &= -\frac{1}{\pi_i}, \qquad i = 1, \dots, n.
\end{aligned} \tag{4.15}$$

These formulae can be obtained either directly, or by using the limit procedure $\pi_{n+1} \to 0$ in (4.14). Also, in this second case, we can show that V is the orthocenter of the simplex.

For the purpose of this section, we shall call an orthocentric simplex different from the right, i.e. of type (i) or (iii) in Theorem 4.2.1 *proper*.

Using the formulae (4.14), we can show easily:

Theorem 4.2.2 *A simplex Σ is proper orthocentric if and only if there exist nonzero numbers $c_1, \dots, c_{n+1}$ such that:*

(i) *either all of them have the same sign, or,*
(ii) *one of them has a sign different from the remaining, and the interior angles φ_{ij} of Σ satisfy*

$$\cos \varphi_{ij} = \varepsilon c_i c_j, \qquad i \neq j,\ i, j = 1, \dots, n+1; \tag{4.16}$$

here, $\varepsilon = 1$ in case (i) *and -1 in case* (ii).

Proof. Let Σ be proper orthocentric. If all the numbers π_i in Theorem 4.2.1 are positive, then, for $i \neq j$, by (4.14)

$$\cos \varphi_{ij} = -\frac{q_{ij}}{\sqrt{q_{ii}}\sqrt{q_{jj}}} = \frac{1}{\pi_i \sqrt{q_{ii}}} \cdot \frac{1}{\pi_j \sqrt{q_{jj}}},$$

so that (4.16) is fulfilled for $c_i = 1/\pi_i \sqrt{q_{ii}} \neq 0$.

If one of the numbers π_i is negative, then ρ in (4.14) is negative and (4.16) is also fulfilled.

Suppose now that (4.16) holds. Then, for $i \neq j$

$$q_{ij} = -\sqrt{q_{ii}}\sqrt{q_{jj}} \cos \varphi_{ij} = -\varepsilon \sqrt{q_{ii}}\sqrt{q_{jj}} c_i c_j = -\varepsilon \lambda_i \lambda_j$$

for $\lambda_i = \sqrt{q_{ii}} c_i \neq 0$, $i = 1, \dots, n+1$. Let us show that the point $V = (\lambda_1, \dots, \lambda_{n+1})$ is the orthocenter of the simplex.

However, this follows immediately from the fact that there is a linear dependence relation $\varepsilon \lambda_k V + S_k = (\varepsilon \lambda_k^2 + q_{kk}) A_k$ between each point A_k, $V = (\lambda_1, \dots, \lambda_{n+1})$, and the direction $S_k = (q_{1k}, \dots, q_{kk}, \dots, q_{k,n+1})$, perpendicular to ω_k. □

Corollary 4.2.3 *The extended signed graph of an orthocentric n-simplex belongs to one of the following three types:*

(i) *A complete positive graph with $n+2$ nodes.*
(ii) *There is a subset S of n nodes among which all edges are negative; the remaining two nodes are connected to all nodes in S by positive edges and the edge between them is negative.*
(iii) *There is a subset S of n nodes among which all edges are missing; the remaining two nodes are connected to all nodes in S by positive edges and the edge between them is negative.*

Before proceeding further, we recall the notion of the *d-rank* of a square matrix A. It was defined in [29] as the number

$$d(A) = \min_D \{\operatorname{rank}(A + D) : D \text{ is a diagonal matrix}\}.$$

In [23], conditions were found for which $d(A^{-1}) = d(A)$ if A is nonsingular.

Theorem 4.2.4 *Let A be an $n \times n$ matrix, $n \geq 3$. If neither of the matrices A and A^{-1} has a zero entry, then A has d-rank one if and only if A^{-1} has d-rank one.*

Proof. Under our assumptions, let $d(A) = 1$. Then $A = D_0 + XY^T$, where both X and Y are $n \times 1$ matrices with no zero entry and D_0 is diagonal. If D_0 were singular, then just one diagonal entry, say the last one, would be zero, because of the rank condition. But then the (1,2)-entry of A^{-1} would be zero (the adjoint of A has two columns proportional), a contradiction. Therefore, D_0 is nonsingular. Observe that the number $1 + Y^T D_0^{-1} X$ is different from zero since $\det A = \det D_0(1 + Y^T D_0^{-1} X)$. Thus

$$A^{-1} = D_0^{-1} - D_0^{-1} X (1 + Y^T D_0^{-1} X)^{-1} Y^T D_0^{-1}$$

so that $d(A^{-1}) = 1$. Because of the symmetry of the formulation with respect to inversion, the converse is also true. The rest is obvious. □

Using the notion of the d-rank, we can prove:

Theorem 4.2.5 *A simplex is proper orthocentric if and only if the Gramian has d-rank one.*

Proof. This follows immediately from Theorem 4.2.2 and (ii) of Theorem 2.1.1. □

Before proceeding further, we shall find other characterizations of the *acute* orthocentric n-simplex.

Theorem 4.2.6 *Let Σ be an n-simplex in $\mathcal{E}_n$ with vertices $A_1, \ldots, A_{n+1}$. Then the following are equivalent:*

(i) *Σ is an acute orthocentric simplex.*
(ii) *In an $(n+1)$-dimensional Euclidean space $\mathcal{E}_{n+1}$ containing $\mathcal{E}_n$, there exists a point P such that the $n+1$ halflines PA_i are mutually perpendicular in $\mathcal{E}_{n+1}$.*
(iii) *There exist $n+1$ hyperspheres in $\mathcal{E}_n$, each with center in A_i, $i = 1, \ldots, n+1$ such that any two of them are orthogonal.*

In case (ii), *the orthogonal projection of P onto $\mathcal{E}_n$ is the orthocenter of Σ. In case* (iii), *the point having the same potence with respect to all $n+1$ hyperspheres is the orthocenter of Σ.*

Proof. (i) $\leftrightarrow$ (ii). Let Σ be an acute orthocentric simplex. Then there exist *positive* numbers $\pi_1, \ldots, \pi_{n+1}$ such that for the squares of the edges, $m_{ij} = \pi_i + \pi_j$, $i \neq j$, $i, j = 1, \ldots, n+1$.

In the space $\mathcal{E}_{n+1}$, choose a perpendicular line L through the orthocenter V of Σ and choose P as such a point of L which has the distance $|\pi_{n+2}|$ from $\mathcal{E}_n$. If i, j are distinct indices from $\{1, \ldots, n+1\}$, the triangle A_iPA_j is by the Pythagorean theorem right with the right angle at P since the square of the distance of P to A_i is π_i.

Conversely, if P is such a point that all the PA_is are perpendicular, choose π_i as the square of the distance of P to A_i. These π_is are positive and by the Pythagorean theorem, the square of the length of the edge A_iA_j is $\pi_i + \pi_j$ whenever $i \neq j$.

(i) $\leftrightarrow$ (iii). Given Σ, choose the hypersphere H_i with center A_i and radius $\sqrt{\pi_i}$, $i = 1, \ldots, n+1$. If $i \neq j$, then K_i and K_j intersect; if X is any point of the intersection, then A_iXA_j is a right triangle with the right angle at X by the Pythagorean theorem, which implies that K_i and K_j intersect orthogonally.

The converse is also evident by choosing the numbers π_i as squares of the radii of the hyperspheres.

The fact that the orthocenter V of Σ has the same potence with respect to the hyperspheres, namely π_{n+2}, is easily established. □

Remark 4.2.7 Because of the fact that the orthocenter has the same potence with respect to all hyperspheres, K_i can also be formulated as follows. There exists a (formally real, cf. Remark 1.4.12) hypersphere with center in V and a purely imaginary radius which is (in a generalized manner) orthogonal to all hyperspheres K_i. This formulation applies also to the case of an *obtuse* orthocentric n-simplex for which a characterization analogous to (iii) can be proved.

Let us find now the squares of the distances of the orthocenter from the vertices in the case that the simplex is proper. The (nonhomogeneous) barycentric coordinates q_i of the vector $\overrightarrow{A_1V}$ are $q_i = (1/\rho\pi_i) - \delta_{1i}$, where $\rho = \sum_{i=1}^{n+1}(1/\pi_i)$ and $\delta_{.}$ is the Kronecker delta, so that

$$\begin{aligned}\rho^2(A_1, V) &= -\frac{1}{2}\sum_{i,k} m_{ik}q_iq_k \\ &= \sum_{i=1}^{n+1}\pi_i q_i^2 \\ &= \sum_i \pi_i\left(\frac{1}{\rho\pi_i} - \delta_{1i}\right)^2 \\ &= \sum_i\left(\frac{1}{\rho^2\pi_i} - 2\frac{\delta_{1i}}{\rho} + \pi_i\delta_{1i}\right) \\ &= \pi_1 - \frac{1}{\rho}.\end{aligned}$$

If we denote

$$-\frac{1}{\rho} = \pi_{n+2}, \tag{4.17}$$

we have

$$\rho^2(A_1, V) = \pi_1 + \pi_{n+2},$$

and for all $k = 1, \ldots, n+1$

$$\rho^2(A_k, V) = \pi_k + \pi_{n+2}. \tag{4.18}$$

This means that if we denote the point V as A_{n+2}, the relations (4.7) hold, by (4.18), for all $i, k = 1, \ldots, n+2$, and, in addition, by (4.17)

$$\sum_{k=1}^{n+2} \frac{1}{\pi_k} = 0. \tag{4.19}$$

The relations (4.7) and the generalized relations (4.18) are symmetric with respect to the indices $1, 2, \ldots, n+2$. Also, the inequalities

$$\sum_{k=1 k\neq i} \frac{1}{\pi_k} \neq 0$$

are fulfilled for all $i = 1, \ldots, n+2$ due to (4.19). Thus the points $A_1, \ldots, A_{n+2}$ form a system of $n+2$ points in $\mathcal{E}_n$ with the property that each of the points of the system is the orthocenter of the simplex generated by the remaining points. Such a system is called an *orthocentric system* in $\mathcal{E}_n$.

As the following theorem shows, the orthocentric system of points in $\mathcal{E}_n$ can be characterized as the maximum system of distinct points in $\mathcal{E}_n$, the mutual distances of which satisfy (4.18).

Theorem 4.2.8 *Let a system of m distinct points $A_1, \ldots, A_m$ in $\mathcal{E}_n$ have the property that there exist numbers $\lambda_1, \ldots, \lambda_m$ such that the squares of the distances of the points A_i satisfy*

$$\rho^2(A_i, A_k) = \lambda_i + \lambda_k, \qquad i \neq k,\ i, k = 1, \ldots, m. \tag{4.20}$$

Then $m \leq n+2$. If $m = n+2$, then the system is orthocentric. If $m < n+2$, the system can be completed to an orthocentric system in $\mathcal{E}_n$, if and only if $\sum_{k=1}^{m} (1/\lambda_k) \neq 0$. If $\sum_{k=1}^{m} (1/\lambda_k) = 0$, then there exists in $\mathcal{E}_n$ an $(m-2)$-dimensional subspace in which the system is orthocentric.

Proof. By (4.20), at most one of the numbers λ_k is not positive. Let, say, $\lambda_1, \ldots, \lambda_{m-1}$ be positive. By Theorem 4.2.1, the points $A_1, \ldots, A_{m-1}$ are linearly independent and, as points in $\mathcal{E}_n$, $m-1 \leq n+1$, or $m \leq n+2$. Suppose now that $m = n+2$. The points $A_1, \ldots, A_{n+2}$ are thus linearly dependent and there exist numbers $\alpha_1, \ldots, \alpha_{n-2}$, not all equal to zero, and such that both $\sum_{i=1}^{n+2} \alpha_i A_i = 0$ and $\sum_{i=1}^{n+2} \alpha_i = 0$ are satisfied. We have necessarily $\lambda_{n+2} < 0$, since otherwise, by Theorem 4.2.1, the points $A_1, \ldots, A_{n+2}$ would be linearly independent, and $\alpha_{n+2} \neq 0$, since otherwise the points $A_1, \ldots, A_{n+1}$ would be linearly dependent.

Now we apply a generalization of Theorem 1.2.4, which will independently be proved as Theorem 5.5.2. Whenever $\sum_{i=1}^{n+2} x_i = 0$, then

$$\sum_{i,k=1}^{n+2} m_{ik}\, x_i\, x_k = -2\sum_{i=1}^{n+2} \lambda_i\, x_i^2 \leq 0$$

i.e.

$$\sum_{i=1}^{n+2} \lambda_i\, x_i^2 \geq 0$$

with equality, if and only if $x_i = \rho\alpha_i$.

In particular, for $x_i = 1/\lambda_i$, $i = 1, \ldots, n+1$, $x_{n+2} = -\sum_{i=1}^{n+1}(1/\lambda_i)$

$$\sum_{i=1}^{n+1} \frac{1}{\lambda_i} - |\lambda_{n+2}|\left(\sum_{i=1}^{n+1} \frac{1}{\lambda_i}\right)^2 \geq 0,$$

i.e.

$$|\lambda_{n+2}| \leq \frac{1}{\sum_{i=1}^{n+1}(1/\lambda_i)}.$$

By Schwarz inequality, we also have for $x_i = \alpha_i$

$$\begin{aligned} |\lambda_{n+2}|\alpha_{n+2}^2 &= \sum_{k=1}^{n+1} \lambda_k \alpha_k^2 \\ &\geq \frac{1}{\sum_{k=1}^{n+1}(1/\lambda_k)}\left(\sum_{k=1}^{n+1} \alpha_k\right)^2 \\ &= \frac{1}{\sum_{k=1}^{n+1}(1/\lambda_k)}\alpha_{n+2}^2, \end{aligned}$$

so that

$$|\lambda_{n+2}| \geq \frac{1}{\sum_{i=1}^{n+1}(1/\lambda_i)}.$$

This shows that

$$-\lambda_{n+2} = \frac{1}{\sum_{k=1}^{n+1}(1/\lambda_k)},$$

or

$$\sum_{i=1}^{n+2} \frac{1}{\lambda_i} = 0.$$

By (4.19), the system of points $A_1, \ldots, A_{n+2}$ is orthocentric. The converse also holds.

Now, let $m < n+2$, and suppose $\sum_{k=1}^{m} (1/\lambda_k) \neq 0$. Then either all the numbers $\lambda_1, \ldots, \lambda_m$ are positive, or just one is negative and $\sum_{k=1}^{m} (1/\lambda_k) < 0$ (otherwise, such a system of points in $\mathcal{E}_n$ would not exist). In both cases, the points $A_1, \ldots, A_m$ are linearly independent and form vertices of a proper orthocentric $(m-1)$-simplex, the vertices of which can be completed by the orthocenter to an orthocentric system in some $(m-1)$-dimensional space, or also by completing by further positive numbers $\lambda_{m-1}, \ldots, \lambda_{n+1}$ (such that $\sum_{k=1}^{n+1} (1/\lambda_k) < 0$, if one of the numbers $\lambda_1, \ldots, \lambda_m$ is negative) to an orthocentric n-simplex contained in an orthocentric system in $\mathcal{E}_n$. If $\sum_{i=1}^{m} (1/\lambda_i) = 0$, the points $A_1, \ldots, A_m$ are linearly dependent (for $x_i = 1/\lambda_i$, we obtain $\sum m_{ik} x_i x_k = 0$) and by the first part they form an orthocentric system in the corresponding $(m-2)$-dimensional space. □

An immediate consequence of Theorems 4.2.1 and 4.2.8 is the following.

Theorem 4.2.9 *Every (at least two-dimensional) face of a proper orthocentric n-simplex is again a proper orthocentric simplex. Every such face of an acute orthocentric simplex is again an acute orthocentric simplex.*

It is well known that a tetrahedron is (proper) orthocentric if and only if one of the following conditions is fulfilled.

(i) Any two of the opposite edges are perpendicular.
(ii) The sums of squares of the lengths of opposite edges are mutually equal.

Let us show that a certain converse of Theorem 4.2.9 holds.

Theorem 4.2.10 *If all three-dimensional faces $A_i, A_{i+1}, A_j, A_{j+1}$, $1 \leq i < j-1 \leq n-1$ of an n-simplex with vertices $A_1, \ldots, A_{n+1}$ are proper orthocentric tetrahedrons, the simplex itself is proper orthocentric.*

Proof. We use induction with respect to the dimension n to prove the assertion. Since for $n = 3$ the assertion is trivial, let $n > 3$ and suppose that the assertion is true for $(n-1)$-simplexes.

Suppose that $A_1, \ldots, A_{n+1}$ are vertices of Σ. By the induction hypothesis, the $(n-1)$-simplex with vertices $A_1, \ldots, A_n$ is orthocentric so that $|A_iA_j|^2 = \pi_i + \pi_j$ for some real $\pi_1, \ldots, \pi_n$, whenever $1 \leq i < j \leq n$ holds. By the assumption, the tetrahedron A_1, A_2, A_n, A_{n+1} is orthocentric so that

$$|A_1A_{n+1}|^2 + |A_2A_n|^2 = |A_1A_n|^2 + |A_2A_{n+1}|^2 \tag{4.21}$$

by (ii) of Theorem 4.2.9. Choose $\pi_{n+1} = |A_1A_{n+1}|^2 - \pi_1$. We shall show that $|A_kA_{n+1}|^2 = \pi_k + \pi_{n+1}$ for all $k = 1, \ldots, n$. This is true for $k = 1$, as well as for $k = 2$ by (4.21). Since the tetrahedron A_2, A_3, A_n, A_{n+1} is also orthocentric, we obtain the result for $k = 3$, etc., up to $k = n$. □

In the following corollary, we call two edges of a simplex *nonneighboring*, if they do not have a vertex in common. Also, two faces are called *complementary*, if the sets of vertices which generate them are complementary, i.e. they are disjoint and their union is the set of all vertices.

Corollary 4.2.11 *Let Σ be an n-simplex, $n \geq 3$. Then the following are equivalent:*

(i) *Σ is proper orthocentric.*
(ii) *Any two nonneighboring edges of Σ are perpendicular.*
(iii) *Every edge of Σ is orthogonal to its complementary face.*

In the next theorem, a well-known property of the triangle is generalized.

Theorem 4.2.12 *The centroid T, the circumcenter S, and the orthocenter V of an orthocentric n-simplex are collinear. In fact, the point T is an interior point of the segment SV and*

$$\frac{ST}{TV} = \frac{1}{2}(n-1).$$

Proof. By Theorem 2.1.1, the numbers q_{0i} are homogeneous barycentric coordinates of the circumcenter S. We distinguish two cases.

Suppose first that the simplex is not right. Then all the numbers π_i from Theorem 4.2.1 are different from zero and the (not homogeneous) barycentric coordinates s_i of the point S are by (4.14) obtained from

$$-2\tau s_i = \frac{n-1}{\pi_i} - \sum_{k=1}^{n+1} \frac{1}{\pi_k},$$

where

$$\tau = \sum_{k=1}^{n+1} \frac{1}{\pi_k}.$$

The (inhomogeneous) barycentric coordinates v_i of the circumcenter V are by Theorem 4.2.1

$$\tau v_i = \frac{1}{\pi_i},$$

and the coordinates t_i of the centroid T satisfy

$$t_i = \frac{1}{n+1}.$$

We have thus

$$-2\tau S = (n-1)\tau V - \tau(n+1)T,$$

or

$$T = \frac{2}{n+1}S + \frac{n-1}{n+1}V.$$

The theorem holds also for the right n-simplex; the proof uses the relations (4.15). □

Remark 4.2.13 The line ST (if $S \not\equiv T$) is called the *Euler line*, analogously to the case of the triangle.

As is well known, the midpoints of the edges and the heels of the altitudes in the triangle are points of a circle, the so-called *Feuerbach circle*. In the simplex, we even have a richer relationship.[1]

Theorem 4.2.14 *Let $m \in \{1, \ldots, n-1\}$. Then the centroids and the circumcenters of all m-dimensional faces of an orthocentric n-simplex belong to the hypersphere K_m*

$$K_m \equiv (m+1)\sum_{i=1}^{n+1} \pi_i x_i^2 - \sum_{i=1}^{n+1} \pi_i x_i \sum_{i=1}^{n+1} x_i = 0.$$

The centers of the hyperspheres K_m are points of the Euler line.

Remark 4.2.15 For $m = 1$, the orthocenter is to be taken as the orthogonal projection of the orthocenter on the corresponding edge.

Proof. Suppose first that the given n-simplex with vertices A_i is not right. It is immediate that whenever M is a subset of $N = \{1, \ldots, n+1\}$ with $m+1$ elements, then the centroid of the m-simplex with vertices A_i having indices in M has homogeneous coordinates $t_i = 1$ for $i \in M$ and $t_i = 0$ for $i \notin M$. Its orthocenter has coordinates $v_i = 1/\pi_i$ for $i \in M$ and $v_i = 0$ for $i \notin M$. The verification that all these points satisfy the equation of K_m is then easy. The case of the right simplex is proved analogously. □

Another notion we shall need is the generalization of a conic in P_2. A *rational normal curve of degree n* in the projective space P_n is the set of points with projective coordinates $(x_1, x_2, \ldots, x_{n+1})$ which satisfy the system of equations

$$x_i = \overset{i}{a}_0 t_1^n + \overset{i}{a}_1 t_1^{n-1} t_2 + \overset{i}{a}_2 t_1^{n-2} t_2^2 + \cdots + \overset{i}{a}_n t_2^n,$$

where (t_1, t_2) is a homogeneous pair of parameters and $[\overset{i}{a}_k]$ a nonsingular fixed matrix (cf. Appendix, Section 5).

[1] See [2].

By a suitable choice of the coordinate system, every real rational normal curve (of degree n) in P_n passing through the basic coordinate points $O_1, O_2, \ldots, O_{n+1}$ and another point $Y = (y_1, y_2, \ldots, y_{n+1})$ can be written in the form[2]

$$x_1 = \frac{y_1}{t - t_1}, \quad x_2 = \frac{y_2}{t - t_2}, \quad \ldots, \quad x_{n+1} = \frac{y_{n+1}}{t - t_{n+1}}. \tag{4.22}$$

We can now generalize another well-known property of the triangle. Let us call an *equilateral n-hyperbola in $\mathcal{E}_n$* such a rational normal curve in $\mathcal{E}_n$, the n asymptotic directions of which are perpendicular.

In addition, we call two such equilateral n-hyperbolas *independent* if both n-tuples of their asymptotic directions are independent in the sense of Theorem A.5.13.

Theorem 4.2.16 *Suppose that a rational normal curve in $\mathcal{E}_n$ contains $n+2$ points of an orthocentric system. Then this curve is an equilateral n-hyperbola.*

Theorem 4.2.17 *Suppose that two independent equilateral n-hyperbolas in $\mathcal{E}_n$ have $n+2$ distinct points in common. Then these $n+2$ points form an orthocentric system in $\mathcal{E}_n$.*

Proof. We shall prove both theorems together. We start with the proof of the second theorem. Suppose there are two independent equiangular n-hyperbolas containing $n+2$ distinct points in $\mathcal{E}_n$. Let $O_1, O_2, \ldots, O_{n+1}$ be some $n+1$ of them; they are necessarily linearly independent since otherwise the first n-hyperbola would have with (any) hyperplane H containing these points at least $n+1$ points in common and would completely belong to H. Let the remaining point have barycentric coordinates $Y = (y_1, y_2, \ldots, y_{n+1})$ with respect to the simplex with basic vertices in $O_1, \ldots, O_{n+1}$. By the same reasoning as above, $y_i \neq 0$, $i = 1, \ldots, n+1$. Denote again by m_{ij} the squares of the distances between the points O_i and O_j.

Both real rational normal curves containing the points $O_1, \ldots, O_{n+1}$, Y, have equations of the form (4.22); the second, say, with numbers $t'_1, t'_2, \ldots, t'_{n+1}$.

By assumption they are independent, thus having both n-tuples of asymptotic directions independent and, by Theorem A.5.13 in the Appendix, there is a unique nonsingular quadric (in the improper hyperplane) with respect to which both n-tuples of directions are autopolar.

One such quadric is the imaginary intersection of the circumscribed hypersphere $\sum m_{ik} x_i x_k = 0$ with the improper hyperplane $\sum x_i = 0$.

[2] Its usual parametric form is obtained by multiplication of the right-hand sides by the product $(t - t_1) \ldots (t - t_{n+1})$.

We now show that the intersection of the improper hyperplane with the quadric $a \equiv \sum_{i,j=1}^{n+1} a_{ij}\, x_i\, x_j = 0$, $a_{ij} = 0$ for $i = j$, $a_{ij} = \dfrac{1}{y_i} + \dfrac{1}{y_j}$ for $i \neq j$, has this property. The first n-hyperbola has the form (4.22); denote by Z_r, $r = 1, \ldots, n$, its improper points, i.e. the points $Z_r = (\overset{r}{z}_1, \overset{r}{z}_2, \ldots, \overset{r}{z}_{n+1})$, where $\overset{r}{z}_i = \dfrac{y_i}{\tau_r - t_i}$, $\tau_1, \ldots, \tau_n$ being the (necessarily distinct) roots of the equation

$$\sum_{i=1}^{n+1} \frac{y_i}{\tau - t_i} = 0. \tag{4.23}$$

However, for $r \neq s$

$$\begin{aligned}
\sum_{i,j=1}^{n+1} a_{ij}\, \overset{r}{z}_i \overset{s}{z}_j &= \sum_{i \neq j, i,j=1}^{n+1} \left(\frac{1}{y_i} + \frac{1}{y_j} \right) \frac{y_i\, y_j}{(\tau_r - t_i)(\tau_s - t_j)} \\
&= \sum_{i,j=1}^{n+1} \frac{y_i + y_j}{(\tau_r - t_i)(\tau_s - t_j)} - 2 \sum_{i=1}^{n+1} \frac{y_i}{(\tau_r - t_i)(\tau_s - t_i)} \\
&= \sum_{i=1}^{n+1} \frac{y_i}{\tau_r - t_i} \sum_{j=1}^{n+1} \frac{1}{\tau_s - t_j} + \sum_{j=1}^{n+1} \frac{y_j}{\tau_s - t_j} \sum_{i=1}^{n+1} \frac{1}{\tau_r - t_i} \\
&\quad - \frac{2}{\tau_s - \tau_r} \left[\sum_{i=1}^{n+1} \frac{y_i}{\tau_r - t_i} - \sum_{i=1}^{n+1} \frac{y_i}{\tau_s - t_i} \right] \\
&= 0
\end{aligned}$$

in view of (4.23); the asymptotic directions Z_r and Z_s are thus conjugate points with respect to the quadric a.

The same is also true for the second hyperbola. This implies that the improper part of the quadric a coincides with the previous. Thus a is a hypersphere, and, since it contains the points $O_1, \ldots, O_{n+1}$, it is the circumscribed hypersphere. Consequently, for some $\rho \neq 0$ and $i \neq j$

$$m_{ij} = \rho \left(\frac{1}{y_i} + \frac{1}{y_j} \right), \tag{4.24}$$

i.e.

$$m_{ij} = \pi_i + \pi_j, \qquad i \neq j.$$

The simplex $O_1, \ldots, O_{n+1}$ is thus orthocentric and the point $Y(1/\pi_i)$ is its orthocenter. Since this orthocentric simplex is not right, the given system of $n + 2$ points is indeed orthocentric.

To prove Theorem 4.2.16, suppose that $O_1, \ldots, O_{n+1}, Y$ is an orthocentric system in $\mathcal{E}_n$. Choose the first $n+1$ of them as basic coordinate points of a simplex and let $Y = (y_i)$ be the last point, so that (4.24) holds. We already saw that every real rational normal curve of degree n, which contains the points

$O_1, O_2, \ldots, O_{n+1}, Y$, has by (4.22) the property that its improper points are conjugate with respect to the quadric $\sum_{i,j=1}^{n+1} m_{ij}\, x_i\, x_j = 0$, i.e. with respect to the circumscribed hypersphere. This means that every such curve is an equilateral n-hyperbola. □

We introduce now the notion of an equilateral quadric and find its relationship to orthocentric simplexes and systems. We start with a definition.

Definition 4.2.18 We call a point quadric α with equation

$$\sum_{i,k=1}^{n+1} \alpha_{ik} x_i x_k = 0$$

in projective coordinates and a dual quadric b with equation

$$\sum_{i,k=1}^{n+1} b_{ik}\, \xi_i\, \xi_k = 0$$

in dual coordinates of the same space *apolar*, if $\sum_{i,k=1}^{n+1} \alpha_{ik}\, b_{ik} = 0$.

Remark 4.2.19 It is well known that apolarity is a geometric notion that does not depend on the coordinate system. One such geometric characterization of apolarity is, in terms of the quadrics α and b:

> There exists a simplex which is autopolar with respect to b and all of whose vertices are points of α.

One direction is easy: in such a case all the b_{ik}s with $i \neq k$ are equal to zero, whereas all the a_{ii}s are equal to zero. The converse is more complicated and we shall not prove it.

Definition 4.2.20 We call a quadric in the Euclidean space *equilateral* if it is apolar to the isotropic improper dual quadric.

Remark 4.2.21 Observe that if the dimension of the space is two, a nonsingular quadric (in this case a conic) is equilateral if and only if it is an equilateral hyperbola.

Remark 4.2.22 In the barycentric coordinates with respect to a usual n-simplex Σ, the condition that the quadric $\sum_{i,k=1}^{n+1} \alpha_{ik}\, x_i\, x_k = 0$ is equilateral is given by

$$\sum_{i,k=1}^{n+1} q_{ik}\, \alpha_{ik} = 0.$$

Theorem 4.2.23 *Suppose that an n-simplex Σ is orthocentric but not a right one. Then every equilateral quadric containing all vertices of Σ contains the orthocenter as well.*

Conversely, every quadric containing all vertices, as well as the orthocenter of Σ, is equilateral.

Proof. Let $\alpha \equiv \sum_{i,k=1}^{n+1} \alpha_{ik}\, x_i\, x_k = 0$ be equilateral, containing all the vertices of Σ. Then

$$\alpha_{ii} = 0, \qquad i = 1, \ldots, n+1, \tag{4.25}$$

$$\sum_{i,k=1}^{n+1} \alpha_{ik}\, q_{ik} = 0. \tag{4.26}$$

By the formulae (4.14), (4.26) can be written as

$$\sum_{i,k=1}^{n+1} \alpha_{ik} \frac{1}{\pi_i} \frac{1}{\pi_k} = 0, \tag{4.27}$$

since by (4.25), the numbers q_{ii} are irrelevant.

This means, however, that α contains the orthocenter.

If, conversely, the quadric α contains all vertices as well as the orthocenter, then both (4.25) and (4.27) hold, thus by (4.14) also (4.26). Consequently, α is equilateral. □

In the sequel, we shall use the theorem:

Theorem 4.2.24 *A real quadric in a Euclidian n-dimensional space is equilateral if and only if it contains n mutually orthogonal asymptotic directions.*

Proof. Suppose first that a real point quadric α contains n orthogonal asymptotic directions. Choose these directions as directions of the axes of some cartesian coordinate system. Then the equation of α in homogeneous cartesian coordinates (with the improper hyperplane $x_{n+1} = 0$) is

$$\sum_{i,k=1}^{n+1} \alpha_{ik}\, x_i\, x_k = 0, \quad \text{where } \alpha_{ii} = 0 \quad \text{for } i = 1, \ldots, n.$$

The isotropic quadric has then the (dual) equation $\xi_1^2 + \cdots + \xi_n^2 = 0 \equiv \sum_{i,k=1}^{n+1} b_{ik}\, \xi_i\, \xi_k = 0$. Since $\sum_{i,k=1}^{n+1} \alpha_{ik}\, b_{ik} (= \sum_{i=1}^{n} \alpha_{ii}) = 0$, α is equilateral.

We shall prove the second part by induction with respect to the dimension n of the space. If $n = 2$, the assertion is correct since the quadric is then an equilateral hyperbola. Thus, let $n > 2$ and suppose the assertion is true for equilateral quadrics of dimension $n - 1$.

We first show that the given quadric α contains at least one asymptotic direction.

In a cartesian system of coordinates in $\overline{\mathcal{E}_n}$, the equation of the dual isotropic quadric is $\sum_{i=1}^{n} \xi_i^2 = 0$, so that for the equilateral quadric $\alpha \equiv \sum_{i,k=1}^{n+1} \alpha_{ik}\, x_i\, x_k = 0$, $\sum_{i=1}^{n} \alpha_{ii} = 0$ holds. If all the numbers α_{ii} are equal to zero, our claim is true (e.g. for the direction $(1, 0, \ldots, 0)$). Otherwise, there exist two numbers, say α_{jj} and α_{kk}, with different signs. The direction $(c_1, \ldots, c_{n+1})$, for which c_j, c_k are the (necessarily real) roots of the equation $\alpha_{jj}\, c_j^2 + 2\alpha_{jk}\, c_j\, c_k + \alpha_{kk}\, c_k^2 = 0$, whereas the remaining c_i are equal to zero, is then a real asymptotic direction of the quadric α.

Thus let s be some real asymptotic direction of α. Choose a cartesian system of coordinates in $\overline{\mathcal{E}_n}$ such that the coordinates of the direction s are $(1, 0, \ldots, 0)$. If the equation of α in the new system is $\sum_{i,k=1}^{n+1} \alpha_{ik}\, x_i\, x_k = 0$, then $\alpha_{11} = 0$. The dual equation of the isotropic quadric is $\xi_1^2 + \cdots + \xi_n^2 = 0$, so that $\sum_{i=1}^{n} \alpha_{ii} = 0$, and thus also $\sum_{i=2}^{n} \alpha_{ii} = 0$. This implies that in the hyperplane $\overline{\mathcal{E}_{n-1}}$ with equation $x_1 = 0$, which is perpendicular to the direction s, the intersection quadric $\tilde{\alpha}$ of the quadric α with $\overline{\mathcal{E}_{n-1}}$ has the equation $\sum_{i,k=2}^{n+1} \alpha_{ik}\, x_i\, x_k = 0$. Since the dual equation of the isotropic quadric in $\mathcal{E}_{n-1}$ is $\xi_2^2 + \cdots + \xi_n^2 = 0$, the quadric $\tilde{\alpha}$ is again equilateral since $\sum_{i=2}^{n} \alpha_{ii} = 0$. By the induction hypothesis, there exist in $\tilde{\alpha}$ $n - 1$ asymptotic directions, which are mutually orthogonal. These form, together with s, n mutually orthogonal directions of the quadric α. □

We now present a general definition.

Definition 4.2.25 A point algebraic manifold ν is called *2-apolar* to a dual algebraic manifold V if the following holds: whenever α is a point quadric containing ν and b is a dual quadric containing V, then α is apolar to b.

In this sense, the following theorem was presented in [13].

Theorem 4.2.26 *The rational normal curve S_n of degree n with parametric equations $x_i = t_1^i\, t_2^{n-i}$, $i = 0, \ldots, n$, in the projective n-dimensional space is 2-apolar to the dual quadric $b \equiv \sum b_{ik}\, \xi_i\, \xi_k = 0$ if and only if the matrix $[b_{ik}]$ is a nonzero Hankel matrix, i.e. if $b_{ik} = c_{i+k}$, $i, k = 0, 1, \ldots, n$, for some numbers $c_0, c_1, \ldots, c_{2n}$, not all equal to zero.*

Proof. Suppose first that S_n is 2-apolar to b. Let i_1, k_1, i_2, k_2 be some indices in $\{0, 1, \ldots, n\}$ such that $i_1 + k_1 = i_2 + k_2$, $i_1 \neq i_2$, $i_1 \neq k_2$. Since S_n is contained in the quadric $x_{i_1}x_{k_1} - x_{i_2}x_{k_2} = 0$, this quadric is apolar to b. Therefore, $b_{i_1\,k_1} = b_{i_2\,k_2}$, so that $[b_{ik}]$ is Hankel.

Conversely, let $[b_{ik}]$ be a Hankel matrix, i.e. $b_{ik} = c_{i+k}$, $i, k = 0, 1, \ldots, n$, and let a point quadric $\alpha \equiv \sum_{i,k=0}^{n} \alpha_{ik}\, x_i\, x_k = 0$ contain S_n. This means that

$$\sum_{i,k=0}^{n} \alpha_{ik}\, t_1^{i+k} t_2^{2n-i-k} \equiv 0.$$

Consequently

$$\sum_{i,k=0,i+k=r}^{n} \alpha_{ik} = 0, \qquad r = 0, \ldots, 2n,$$

so that $\sum_{i,k=0}^{n} \alpha_{ik}\, b_{ik} = \sum_{r=0}^{n} \sum_{i,k=0,i+k=r}^{n} \alpha_{ik}\, b_{ik} = \sum_{r=0}^{n} c_r \sum_{i+k=r}^{n} \alpha_{ik} = 0$. It follows that α is apolar to b, and thus S_n is 2-apolar to b, as asserted. □

We shall use the following classical theorem.

Theorem 4.2.27 *A real positive semidefinite Hankel matrix of rank r can be expressed as a sum of r positive semidefinite Hankel matrices of rank one.*

We are now able to prove:

Theorem 4.2.28 *A rational normal curve of degree n in a Euclidean n-space $\mathcal{E}_n$ is 2-apolar to the dual isotropic quadric of $\overline{\mathcal{E}_n}$ if and only if it is an equilateral n-hyperbola.*

Proof. An equilateral n-hyperbola H has n asymptotic directions mutually perpendicular. Thus every quadric that contains H contains this n-tuple as well. By Theorem 4.2.24, this quadric is an equilateral quadric; thus, by Theorem 4.2.20, apolar to the isotropic quadric. Therefore, every equilateral n-hyperbola is 2-apolar to the isotropic quadric.

Conversely, suppose that S_n is a rational normal curve in a Euclidean n-space $\mathcal{E}_n$, which is 2-apolar to the isotropic quadric of $\mathcal{E}_n$. There exists a coordinate system in $\overline{\mathcal{E}_n}$ in which S_n has parametric equations $x_i = t_1^i t_2^{n-i}$, $i = 0, \ldots, n$. The isotropic quadric b has then an equation, the coefficients of which form, by Theorem 4.2.26, an $(n+1) \times (n+1)$ Hankel matrix B. This matrix is positive semidefinite of rank n. By Theorem 4.2.27, B is a sum of n positive semidefinite Hankel matrices of rank one, $B = \sum_{j=1}^{n} B_j$. Every Hankel positive semidefinite matrix of rank one has the form $[p_{ik}]$, $p_{ik} = y^{i+k} z^{2n-i-k}$, where (y, z) is a real nonzero pair. Hence b has equation

$$\sum_{j=1}^{n} (z_j^n \xi_0 + y_j z_j^{n-1} \xi_1 + y_j^2 z_j^{n-2} \xi_2 + \cdots + y_j^n \xi_n)^2,$$

which implies that the n-tuples $(z_j^n, y_j z_j^{n-1}, \ldots, y_j^n)$ of points of S_n form an autopolar $(n-1)$-simplex of the quadric b. These n points are improper points of $\overline{\mathcal{E}_n}$ (since b is isotropic), and, being asymptotic directions of S_n, are thus perpendicular. □

In the conclusion, we investigate finite sets of points, which are 2-apolar to the isotropic quadric.

Definition 4.2.29 A *generalized orthocentric system* in a Euclidean n-space $\mathcal{E}_n$ is a system of any $m \leq 2n$ mutually distinct points in $\overline{\mathcal{E}_n}$, which (as a point manifold) is 2-apolar to the dual isotropic quadric of $\overline{\mathcal{E}_n}$.

Theorem 4.2.30 *A system of m points $\overset{r}{a} \equiv (\overset{r}{a}_1, \ldots, \overset{r}{a}_{n+1})$, $r = 1, \ldots, m$, is 2-apolar to the dual quadric b (in a projective n-space) if and only if b has the form*

$$b \equiv \sum_{r=1}^{m} \lambda_r \left(\sum_{i=1}^{n+1} \overset{r}{a}_i \, \xi_i \right)^2 = 0 \tag{4.28}$$

for some $\lambda_1, \ldots, \lambda_m$.

Proof. Suppose b has the form (4.28), so that $b_{ik} = \sum_{r=1}^{m} \lambda_r \overset{r}{a}_i \overset{r}{a}_k$, $i, k = 1, \ldots, n+1$. If $\alpha \equiv \sum_{i,k=1}^{n-1} \alpha_{ik} \, x_i \, x_k = 0$ is a quadric containing all the points $\overset{r}{a}$, then $\sum_{i,k} \alpha_{ik} \overset{r}{a}_i \overset{r}{a}_k = 0$ for $r = 1, \ldots, m$. We have thus also $\sum_{i,k=1}^{n+1} \alpha_{ik} b_{ik} = 0$, i.e. α is apolar to b.

Conversely, suppose that whenever α is a quadric containing all the points $\overset{r}{a}$, then α is apolar to $b \equiv \sum b_{ik} \, \xi_i \, \xi_k$. This means that whenever

$$\sum_{i,k=1}^{n+1} \alpha_{ik} \overset{r}{a}_i \overset{r}{a}_k = 0, \ \ \alpha_{ik} = \alpha_{ki}, \ \ r = 1, \ldots, m,$$

holds, then also

$$\sum_{i,k=1}^{n+1} \alpha_{ik} \, b_{ik} = 0$$

holds. Since the conditions are linear, we obtain that

$$b_{ik} = \sum_{r=1}^{m} \lambda_r \overset{r}{a}_i \overset{r}{a}_k, \qquad i, k = 1, \ldots, n+1.$$

This implies (4.28). □

Theorem 4.2.31 *An orthocentric system with $n+2$ points in a Euclidean space $\mathcal{E}_n$ is at the same time a generalized orthocentric system in $\mathcal{E}_n$.*

Proof. Choose $n+1$ of these points as vertices of an n-simplex Σ, so that the remaining point is the orthocenter of this simplex. In our usual notation, the orthocenter has barycentric coordinates $(1/\pi_i)$, and by equations (4.15), we have identically

$$\rho \sum_{i,k=1}^{n+1} q_{ik}\, \xi_i\, \xi_k \equiv \sum_{r=1}^{n+1} \left(\sum_{k=1}^{n+1} \frac{1}{\pi_k} \right) \frac{\xi_r^2}{\pi_r} - \left(\sum_{i=1}^{n+1} \frac{\xi_i}{\pi_i} \right)^2 .$$

By Theorem 4.2.30, the given points form a system of $n+2$ ($\leq 2n$) points 2-apolar to the isotropic quadric. □

In particular, generalized orthocentric systems in $\mathcal{E}_n$, which consist of $2n$ points, are interesting. They can be obtained (however, not all of them) in the way presented in the following theorem:

Theorem 4.2.32 *Let S be an equilateral n-hyperbola and let Q be any equilateral quadric which does not contain S, but intersects S in $2n$ distinct real points. In such a case, these $2n$ points form a generalized orthocentric system.*

Proof. Suppose that R is an arbitrary quadric in $\mathcal{E}_n$ which contains those $2n$ intersection points. If $R \equiv Q$, then R is equilateral. If $R \not\equiv Q$, choose on S an arbitrary point p different from all the intersection points. Since $p \notin Q$, there exists in the pencil of quadrics $\alpha Q + \beta R$ a quadric P containing the point p. The quadric P has with the curve S at least $2n+1$ points in common. By a well-known result from basic algebraic geometry (since S is irreducible of degree n and P is of degree 2), P necessarily contains the whole curve S and is thus equilateral.

The number $\beta \neq 0$ since $p \notin Q$. Consequently, the quadric R is also equilateral. Thus the given system is generalized orthocentric as asserted. □

Remark 4.2.33 The quadric consisting of two mutually distinct hyperplanes is clearly equilateral if and only if these hyperplanes are orthogonal. We can thus choose as Q in Theorem 4.2.23 a pair of orthogonal hyperplanes.

Theorem 4.2.34 *If $2n$ points form a generalized orthocentric system in $\mathcal{E}_n$, then whenever we split these points into two subsystems of n points each, the two hyperplanes containing the systems are perpendicular.*

Example 4.2.35 Probably the simplest example of $2n$ points forming a generalized orthocentric system is the following: let $a_1, \dots, a_n, b_1, \dots, b_n$ be real numbers, all different from zero, $a_i \neq b_i$, $i = 1, \dots, n$, and such that

$$\sum_{i=1}^{n} \frac{1}{a_i b_i} = 0.$$

Then the $2n$ points in an n-dimensional Euclidean space $\mathcal{E}_n$ with a cartesian coordinate system $A_1 = (a_1, 0, \dots, 0)$, $A_2 = (0, a_2, 0, \dots, 0)$, …, $A_n = (0, \dots,$

$0, a_n)$, $B_1 = (b_1, 0, \ldots, 0)$, $B_2 = (0, b_2, 0, \ldots, 0)$, $\ldots$, $B_n = (0, \ldots, 0, b_n)$ form a generalized orthocentric system.

Indeed, choosing for $i = 1, \ldots, n$, $\alpha_i = \frac{1}{a_i(a_i - b_i)}$, $\beta_i = \frac{1}{b_i(b_i - a_i)}$, then for the points $\overset{i}{a} = (0, \ldots, 0, a_i, 0, \ldots, 0, 1)$, and $\overset{i}{b} = (0, \ldots, 0, b_i, 0, \ldots, 0, 1)$ in the projective completion of $\mathcal{E}_n$, (4.28) reads

$$\sum_{i=1}^{n} \alpha_i (a_i \xi_i + \xi_{n+1})^2 + \sum_{i=1}^{n} \beta_i (b_i \xi_i + \xi_{n+1})^2 = \sum_{i=1}^{n} \xi_i^2,$$

which is the equation of the isotropic quadric.

4.3 Cyclic simplexes

In this section, the so-called *normal polygons* in the Euclidean space play the crucial role.

Definition 4.3.1 Let $\{A_1, A_2, \ldots, A_{n+1}\}$ be a cyclically ordered set of $n+1$ linearly independent points in a Euclidean n-space $\mathcal{E}_n$. We call the set of these points, together with the set of $n+1$ segments A_1A_2, A_2A_3, $A_3A_4, \ldots, A_{n+1}A_1$ a *normal polygon* in $\mathcal{E}_n$; we denote it as $V = [A_1, A_2, \ldots, A_{n+1}]$.

The points A_i are the *vertices*, and the segments A_iA_{i+1} are the *edges* of the polygon V.

It is clear that we can assign to every normal $(n+1)$-gon $[A_1, A_2, \ldots, A_{n+1}]$ a cyclically ordered set of $n+1$ vectors $\{v_1, v_2, \ldots, v_{n+1}\}$ such that $v_i = \overrightarrow{A_iA_{i+1}}, i = 1, 2, \ldots, n+1$ (and $A_{n+2} = A_1$). Then $\sum_{i=1}^{n+1} v_i = 0$, and any arbitrary m ($m < n+1$) of the vectors $v_1, v_2, \ldots, v_{n+1}$ are linearly independent. If conversely $\{v_1, v_2, \ldots, v_{n+1}\}$ form a cyclically ordered set of $n+1$ vectors in $\mathcal{E}_n$ for which $\sum_{i=1}^{n+1} v_i = 0$ holds and if any $m < n+1$ vectors of these are linearly independent, then there exists in $\mathcal{E}_n$ a normal polygon $[A_1, A_2, \ldots, A_{n+1}]$ such that $v_i = \overrightarrow{A_iA_{i+1}}$ holds. It is also evident that choosing one of the vertices of a normal polygon as the first (say, A_1), then the Gram matrix $M = [\langle v_i, v_j \rangle]$ of these vectors has a characteristic property that it is positive semidefinite of order $n+1$ and rank n, satisfying $Me = 0$, where e is the vector of all ones.

Observe that also conversely such a matrix determines a normal polygon in $\mathcal{E}_n$, even uniquely up to its position in the space and the choice of the first vertex.

To simplify formulations, we call the following vertex the second vertex, etc., and use again the symbol $V = [A_1, A_2, \ldots, A_{n+1}]$. All the definitions and theorems can, of course, be formulated independently of this choice.

Definition 4.3.2 Suppose that $V_1 = [A_1, A_2, \ldots, A_{n+1}]$ and $V_2 = [B_1, B_2, \ldots, B_{n+1}]$ are two normal polygons in $\mathcal{E}_n$. We call the polygon V_2 *left*(respectively, *right*) *conjugate* to the polygon V_1, if for $k = 1, 2, \ldots, n+1$ (and $A_{n+2} = A_1$) the line $A_k A_{k+1}$ is perpendicular to the hyperplane β_k (respectively, β_{k+1}), where β_i is the hyperplane determined by the points $A_1, \ldots, A_{i-1}, A_{i+1}, \ldots, A_{n+1}$.

Theorem 4.3.3 *Let V_1 and V_2 be normal polygons in $\mathcal{E}_n$. If V_2 is left (respectively, right) conjugate to V_1, then V_1 is right (respectively, left) conjugate to V_2.*

Proof. Suppose that $V_2 = [B_1, B_2, \ldots, B_{n+1}]$ is left conjugate to $V_1 = [A_1, A_2, \ldots, A_{n+1}]$, so that the line $A_k A_{k+1}$ is perpendicular to the lines $B_{k+1}B_{k+2}$, $B_{k+2}B_{k+3}$, $\ldots$, $B_{n+1}B_1$, B_1B_2, $\ldots$, $B_{k-2}B_{k-1}$. It follows that for $k = 1, 2, \ldots, n+1$ (with $B_{n+2} = B_1$) the line $B_k B_{k+1}$ is perpendicular to the lines $A_k A_{k-1}$, $A_{k-1}A_{k-2}$, $\ldots$, $A_1 A_{n+1}$, $A_{n+1}A_n$, $\ldots$, $A_{k+3}A_{k+2}$, thus also to the hyperplane α_{k+1} determined by the points $A_1, \ldots, A_k, A_{k+2}, \ldots, A_{n+1}$. Therefore, V_1 is right conjugate to V_2. The second case can be proved analogously. □

Theorem 4.3.4 *To every normal polygon V_1 in $\mathcal{E}_n$ there exists in $\mathcal{E}_n$ a normal polygon V_2 (respectively, V_3), which is with respect to it left (respectively, right) conjugate. If V and V' are two normal polygons which are both (left or right) conjugate to V_1, then V and V' are homothetic, i.e. the corresponding edges are parallel and their lengths proportional. The vectors of the edges are either all oriented the same way, or all the opposite way.*

Proof. Suppose $V_1 = [A_1, A_2, \ldots, A_{n+1}]$. Denote by ω_i $(i = 1, 2, \ldots, n+1)$ the hyperplane in $\mathcal{E}_n$ which contains the point A_i and is perpendicular to the line $A_i A_{i+1}$ (again $A_{n+2} = A_1$). Observe that these hyperplanes $\omega_1, \omega_2, \ldots, \omega_{n+1}$ do not have a point in common: if P were to be such a point, then we would have $PA_i < PA_{i+1}$ for $i = 1, 2, \ldots, n+1$ since $P \in \omega_i$ and A_i is the heel of the perpendicular line from A_{i+1} on ω_i. (Also, the hyperplanes $\omega_1, \omega_2, \ldots, \omega_{n+1}$ do not have a common direction since otherwise the points $A_1, A_2, \ldots, A_{n+1}$ would be in a hyperplane.) It follows that the points $B_i = \bigcap_{k=1, k\neq i}^{n+1} \omega_k$, $i = 1, 2, \ldots, n+1$, are linearly independent. It is clear that the normal polygon $V_2 = [B_1, B_2, \ldots, B_{n+1}]$ (respectively, the normal polygon $V_3 = [B_{n+1}, B_1, \ldots, B_n]$) is left (respectively, right) conjugate to V_1.

If now $V = [B_1, B_2, \ldots, B_{n+1}]$ and $V' = [B'_1, B'_2, \ldots, B'_{n+1}]$ are two normal polygons, both left conjugate to V_1, then we have, by Theorem 4.3.2, that both vectors $v_i = \overrightarrow{B_i B_{i+1}}$ and $v'_i = \overrightarrow{B'_i B'}_{i+1}$ of the corresponding edges are perpendicular to the same hyperplane and thus parallel, and, in addition, they sum to the zero vector. This implies that for some nonzero constants $c_1, c_2, \ldots, c_{n+1}$, $v_i' = c_i v_i$. Thus $\sum_{i=1}^{n+1} c_i v_i = 0$; since $\sum_{i=1}^{n+1} v_i = 0$ is the only linear relation among $v_1, \ldots, v_{n+1}$, we finally obtain $c_i = C$ for $i = 1, 2, \ldots, n+1$. □

Theorem 4.3.5 *Suppose Σ is an n-simplex, and $v_1, v_2, \ldots, v_{n+1}$ is a system of $n+1$ nonzero vectors, each perpendicular to one $(n-1)$-dimensional face of Σ. Then there exist positive numbers $\alpha_1, \ldots, \alpha_{n+1}$ such that*

$$\sum_{i=1}^{n+1} \alpha_i v_i = 0,$$

if and only if either all the vectors v_i are vectors of outer normals of Σ, or all the vectors v_i are vectors of interior normals of Σ.

Proof. This follows from Theorem 1.3.1 and the fact that the system of vectors $v_1, v_2, \ldots, v_{n+1}$ contains n linearly independent vectors. □

Theorem 4.3.6 *Suppose $V_1 = [A_1, A_2, \ldots, A_{n+1}]$ is a normal polygon in $\mathcal{E}_n$ and $V_2 = [B_1, B_2, \ldots, B_{n+1}]$ a normal polygon left (respectively, right) conjugate to V_1. Then all the vectors $v_i = \overrightarrow{B_i B_{i+1}}$ are vectors of either outer or inner normals to the $(n-1)$-dimensional faces of the simplex determined by the vertices of the polygon V_1.*

Proof. This follows from Theorem 4.3.5 since $\sum_{i=1}^{n+1} v_i = 0$. □

Theorem 4.3.7 *Suppose $V = [A_1, A_2, \ldots, A_{n+1}]$ is a normal polygon in $\mathcal{E}_n$. Assign to every point X in $\mathcal{E}_n$ the sum of squares $\sum_{i=1}^{n+1} \overline{X_i A_i}^2$, where X_i is the heel of the perpendicular from the point X on the line $A_i A_{i+1}$ ($\overline{X_i A_i}$ meaning the distance between X_i and A_i). Then this sum is minimal if X is the center of the hypersphere containing all the points $A_1, A_2, \ldots, A_{n+1}$.*

Proof. Consider the triangle $A_i\ A_{i+1}\ X$. Observe that $\overline{X_i A_i}^2 - \overline{X_i A_{i+1}}^2 = \overline{X A_i}^2 - \overline{X A_{i+1}}^2$. On the other hand, we can see easily from (2.1) that

$$\begin{aligned} &4\overline{A_i A_{i+1}}^2 \overline{X_i A_i}^2 \\ &\quad = (\overline{A_i X_i}^2 - \overline{A_{i+1} X_i}^2)^2 - 2(\overline{A_i X_i}^2 - \overline{A_{i+1} X_i}^2)\overline{A_i A_{i+1}}^2 + \overline{A_i A_{i+1}}^4, \end{aligned}$$

which implies

$$\overline{X_i A_i}^2 = \frac{1}{4}\overline{A_i A_{i+1}}^2 + \frac{1}{2}(\overline{A_i X_i}^2 - \overline{A_{i+1} X_i}^2) + \frac{1}{4\overline{A_i A_{i+1}}^2}(\overline{A_i X_i}^2 - \overline{A_{i+1} X_i}^2)^2.$$

Thus also

$$\overline{X_iA_i}^2 = \frac{1}{4}\overline{A_iA_{i+1}}^2 + \frac{1}{2}(\overline{A_iX}^2 - \overline{A_{i+1}X}^2) + \frac{1}{4}\overline{A_iA_{i+1}}^2(\overline{A_iX}^2 - \overline{A_{i+1}X}^2)^2.$$

Summing these relations for $i = 1, 2, \ldots, n+1$, we arrive at

$$\sum_{i=1}^{n+1} \overline{X_iA_i}^2 = \frac{1}{4}\sum_{i=1}^{n+1} \overline{A_iA_{i+1}}^2 + \sum_{i=1}^{n+1} \frac{1}{4}\overline{A_iA_{i+1}}^2(\overline{A_iX}^2 - \overline{A_{i+1}X}^2)^2,$$

which proves the theorem. □

Theorem 4.3.8 *Suppose Σ is an n-simplex. Let a cyclic (oriented) ordering of all its vertices P_i (and thus also of the opposite $(n-1)$-dimensional faces ω_i) be given, say $P_1, P_2, \ldots, P_{n+1}, P_1$.*

Then there exists a unique normal polygon $V = [A_1, A_2, \ldots, A_{n+1}]$ such that for $i = 1, 2, \ldots, n+1$, the points A_i belong to the hyperplane ω_i and the line A_iA_{i+1} is perpendicular to ω_i. In addition, the circumcenter (in a clear sense) of the polygon V coincides with the Lemoine point of Σ.

If we form in the same way a polygon V' choosing another cyclic ordering of the vertices of Σ, then V' is formed by the same vectors as V, which, however, can have the opposite orientation.

Before we prove this theorem, we present a definition and a remark.

Definition 4.3.9 In the situation described in the theorem, we say that the polygon V is *perpendicularly inscribed* into the simplex Σ.

Remark 4.3.10 Theorem 4.3.8 can also be formulated in such a way that the edges of every such inscribed polygon are segments which join, in the perpendicular way, the two corresponding $(n-1)$-dimensional faces of Σ and the simplex Σ', obtained from Σ by symmetry with respect to the Lemoine point.

Proof. (Theorem 4.3.8) The first part follows immediately from Theorem 4.3.4. The second part is a consequence of the well-known property of the Lemoine point (see Theorem 2.2.3) and the fact that in Theorem 4.3.7, $\overline{A_iX}^2$ is at the same time the square of the distance of X from the hyperplane ω_i.

The third part follows from the second since the length of the vectors of the polygon V, which are perpendicular to ω_i, is twice the distance of the Lemoine point from ω_i. Of course, there are two possible orientations of vectors in V by Theorem 4.3.5. □

Theorem 4.3.11 *Suppose $V_1 = [A_1, A_2, \ldots, A_{n+1}]$ is a normal polygon, and $V_2 = [B_1, B_2, \ldots, B_{n+1}]$ a polygon left (respectively, right) conjugate to V_1. If ω_i are the $(n-1)$-dimensional faces of the n-simplex Σ determined by vertices*

A_i (ω_i opposite to A_i), then the interior angles φ_{ij} of the faces ω_i and ω_j ($i \neq j$) satisfy

$$\cos \varphi_{ij} = \frac{\langle v_i, v_j \rangle}{\sqrt{\langle v_i, v_i \rangle}\sqrt{\langle v_j, v_j \rangle}},$$

where $v_i = \overrightarrow{B_{i-1}B_i}$ (respectively, $v_i = \overrightarrow{B_i B_{i+1}}$).

Proof. By Theorems 4.3.3 and 4.3.2, the vectors v_i are perpendicular to ω_i. Theorem 4.3.6 then implies that the angle between the vectors v_i and v_j ($i \neq j$) is equal to $\pi - \varphi_{ij}$. □

Theorem 4.3.12 *Suppose that $V_1 = [A_1, A_2, \ldots, A_{n+1}]$ and $V_2 = [B_1, B_2, \ldots, B_{n+1}]$ are normal polygons in $\mathcal{E}_n$ such that V_2 is right conjugate to V_1.*

Denote by $a_i = \overrightarrow{A_i A_{i+1}}$, $b_i = \overrightarrow{B_i B_{i+1}}$, $i = 1, 2, \ldots, n+1$; $A_{n+2} = A_1$; $B_{n+2} = B_1$ the vectors of the edges and by $A = [\langle a_i, a_j \rangle]$, $B = [\langle b_i, b_j \rangle]$ the corresponding Gram matrices. Let Z be the $(n+1) \times (n+1)$ matrix

$$Z = \begin{bmatrix} 1 & -1 & 0 & \ldots & 0 \\ 0 & 1 & -1 & \ldots & 0 \\ 0 & 0 & 1 & \ldots & 0 \\ \ldots & \ldots & \ldots & \ldots & \ldots \\ -1 & 0 & 0 & \ldots & 1 \end{bmatrix}. \tag{4.29}$$

Then there exists a nonzero number c such that the matrix

$$\begin{bmatrix} A & cZ \\ cZ^T & B \end{bmatrix} \tag{4.30}$$

is symmetric positive semidefinite of rank n.

Conversely, let (4.30) be a symmetric positive semidefinite matrix of rank n for some number $c \neq 0$. Then A is the Gram matrix of vectors of edges of some normal polygon $V_1 = [A_1, A_2, \ldots, A_{n+1}]$ in some $\mathcal{E}_n$, B is the Gram matrix of vectors of edges of some normal polygon $V_2 = [B_1, B_2, \ldots, B_{n+1}]$ in $\mathcal{E}_n$, and, in addition, V_2 is the right conjugate to V_1.

Proof. Suppose that V_2 is the right conjugate to V_1. Then the vectors a_i, b_i satisfy

$$\langle a_i, b_j \rangle = 0 \tag{4.31}$$

for $i \neq j \neq i+1$. Denote $\langle a_i, b_i \rangle = c_i$, $\langle a_i, b_{i+1} \rangle = d_i$ ($i = 1, \ldots, n+1$, $b_{n+2} = b_1$).

By (4.31) and $\sum a_i = \sum b_i = 0$, we have $c_i = c$, $d_i = -c$, $c \neq 0$ for $i = 1, 2, \ldots, n+1$. It follows that (4.30) is the Gram matrix of the system $a_1, \ldots, a_{n+1}, b_1, \ldots, b_{n+1}$, thus symmetric positive semidefinite of rank n.

Conversely, let (4.30) be a positive semidefinite matrix of rank n for some c different from zero. Let $a_1, a_2, \ldots, a_{n+1}, b_1, b_2, \ldots, b_{n+1}$ be a system of vectors in some n-dimensional Euclidean space $\mathcal{E}_n$, the Gram matrix of which is the matrix (4.30). Since Z has rank n, and $Ze = 0$ for e with all coordinates one, A has also rank n and $Ae = 0$. Similarly, B has rank n and $Be = 0$. Therefore, $\sum a_i = \sum b_i = 0$, and the vectors $a_1, a_2, \ldots, a_{n+1}$ can be considered as vectors of edges, $a_i = \overrightarrow{A_iA_{i+1}}$, of some normal polygon $V_1 = [A_1, A_2, \ldots, A_{n+1}]$, and $b_1, b_2, \ldots, b_{n+1}$ as vectors of edges, $b_i = \overrightarrow{B_iB_{i+1}}$ of some normal polygon $V_2 = [B_1, B_2, \ldots, B_{n+1}]$. Since $\langle a_i, b_j \rangle = 0$ for $i \neq j \neq i+1$, V_2 is the right conjugate to V_1. □

Theorem 4.3.13 *To every symmetric positive semidefinite matrix A of order $n+1$ which has rank n and for which $Ae = 0$, there exists a unique matrix B such that the matrix (4.30) is positive semidefinite and has for a fixed $c \neq 0$ rank n. When $c \neq 0$ is not prescribed, the matrix B is determined uniquely up to a positive multiplicative factor.*

Proof. This follows from Theorems 4.3.12 and 4.3.4. □

Definition 4.3.14 A normal polygon in $\mathcal{E}_n$ is called *orthocentric*, if the simplex with the same vertices is orthocentric.

Theorem 4.3.15 *A normal polygon $V_1 = [A_1, A_2, \ldots, A_{n+1}]$ is orthocentric if and only if the vectors $a_i = \overrightarrow{A_iA_{i+1}}$ satisfy*

$$\langle a_i, a_j \rangle = 0$$

for $j \not\equiv i-1,\ j \not\equiv i, j \not\equiv i+1 \mod (n+1),\ i, j = 1, \ldots, n+1\ (a_{n+2} = a_1)$. The corresponding n-simplex is acute if all the numbers

$$d_i = -\langle a_{i-1}, a_i \rangle,\ i = 1, \ldots, n+1,$$

are positive; it is right (respectively, obtuse) if $d_k = 0$ (respectively, $d_k < 0$) for some index k. In fact, $d_k \leq 0$ cannot occur for more than one k.

Proof. Suppose V_1 is an orthocentric normal polygon, denote by O the orthocenter. Then the vectors $v_i = \overrightarrow{OA_i}$ satisfy

$$\langle v_i, v_j - v_k \rangle = 0$$

for $j \neq i \neq k$. Thus, for $j \not\equiv i-1,\ j \not\equiv i, j \not\equiv i+1 \mod (n+1)$

$$\langle a_i, a_j \rangle = \langle v_{i+1} - v_i, v_{j+1} - v_j \rangle = 0.$$

Conversely, let

$$\langle a_i, a_j \rangle = 0 \qquad \text{for } j \not\equiv i-1,\ j \not\equiv i,\ j \not\equiv i+1 \mod (n+1).$$

If we denote, in addition, by c_i the inner product $\langle a_i, a_i\rangle$, we obtain

$$0 = \left\langle a_i, \sum_1^{n+1} a_j \right\rangle = -d_i + c_i - d_{i+1},$$

thus

$$c_i = d_i + d_{i+1}.$$

A simple computation shows, if $i < k$, that

$$\begin{aligned} 0 &< \langle a_i + a_{i+1} + \cdots + a_{k-1}, a_i + a_{i+1} + \ldots a_{k-1}\rangle \\ &= -\langle a_i + a_{i+1} + \cdots + a_{k-1}, a_k + a_{k+1} + \cdots + a_{n-1} + a_1 + \cdots + a_{i-1}\rangle \\ &= d_i + d_k. \end{aligned}$$

It follows that for at most one k we have $d_k \leq 0$. If $d_k = 0$, the vertex A_k is the orthocenter and the line $A_k A_i$ is perpendicular to $A_k A_j$ for all $i, j, i \neq k \neq j \neq i$ (the simplex is thus right orthocentric): any two of the vectors

$$\begin{aligned} &a_{k-1}, a_k, a_k + a_{k+1} \, (= -a_{k+2} + \cdots + a_{n+1} + a_1 + \cdots + a_{k-1}), \\ &a_k + a_{k+1} + a_{k+2} \, (= -(a_{k+3} + \cdots + a_{n+1} + a_1 + \cdots + a_{k-1})), \\ &\qquad\vdots \\ &a_k + a_{k+1} + \cdots + a_{n+1} + a_1 + \cdots + a_{k-3} (= -(a_{k-2} + a_{k-1})) \end{aligned}$$

are perpendicular.

Suppose now that all the numbers d_k are different from zero. Let us show that the number

$$\gamma = \sum_{k=1}^{n+1} \frac{1}{d_k}$$

is also different from zero, and the point

$$O = \sum_{k=1}^{n+1} \frac{1}{\gamma d_k} A_k$$

is the orthocenter.

Let $i, j = 1, 2, \ldots, n$. Then

$$\begin{aligned} 0 &< \det[\langle a_i, a_j\rangle] \\ &= \det \begin{bmatrix} d_1 + d_2 & -d_2 & 0 & \ldots & 0 & 0 \\ -d_2 & d_2 + d_3 & -d_3 & \ldots & 0 & 0 \\ 0 & -d_3 & d_3 + d_4 & \ldots & 0 & 0 \\ \ldots & \ldots & \ldots & \ldots & \ldots & \ldots \\ 0 & 0 & 0 & \ldots & -d_n & d_n + d_{n+1} \end{bmatrix} \end{aligned}$$

$$= \sum_{i=1}^{n+1} \prod_{j \neq i} d_j = \gamma \prod_{j=1}^{n+1} d_j. \tag{4.32}$$

Since $\sum a_k = 0$, the vectors

$$\begin{aligned} v_i = \overrightarrow{OA_i} &= -\sum_{j=1}^{n+1} \frac{1}{\gamma d_j} \overrightarrow{A_i A_j} \\ &= \frac{1}{\gamma} \left[\frac{1}{d_{i+1}} a_i + \frac{1}{d_{i+2}} (a_i + a_{i+1}) + \dots + \frac{1}{d_{n+1}} (a_i + \dots + a_n) \right. \\ &\quad + \frac{1}{d_1} (a_i + \dots + a_{n+1}) + \dots \\ &\quad \left. + \frac{1}{d_{i-1}} (a_i + \dots + a_{n+1} + a_1 + \dots + a_{i-1}) \right], \\ &\qquad i = 1, \dots, n+1, \end{aligned}$$

satisfy

$$\langle v_i, a_j \rangle = 0$$

for $i - 1 \neq j \neq i$.

If $d_k < 0$ for some k, then, by (4.32), $\gamma < 0$, so that the point O is an exterior point of the corresponding simplex, and due to Section 2 of this chapter, the simplex is obtuse orthocentric. If all the d_ks are positive, $\gamma > 0$ and O is an interior point of the (necessarily acute) simplex. □

Definition 4.3.16 We call an n-simplex Σ ($n \geq 2$) *cyclic*, if there exists such a cyclic ordering of its $(n-1)$-dimensional faces, in which any two not neighboring faces are perpendicular. If then all the interior angles between (in the ordering) neighboring faces are acute, we call the simplex *acutely cyclic*; if one of them is obtuse, we call Σ *obtusely cyclic*. In the remaining case that one of the angles is right, Σ is called *right cyclic*. Analogously, we call also cyclic the normal polygon formed by vertices and edges opposite to the neighboring faces of the cyclic simplex (again acutely, obtusely, or right cyclic).

Remark 4.3.17 The signed graph of the cyclic n-simplex is thus either a positive circuit in the case of the acutely cyclic simplex, or a circuit with one edge negative and the remaining positive in the case of the obtusely cyclic simplex, or finally a positive path in the case of the right cyclic simplex (it is thus a Schlaefli simplex, cf. Theorem 4.1.7).

Theorem 4.3.18 *A normal polygon is an acutely (respectively, obtusely, or right) cyclic if and only if the left or right conjugate normal polygon is acute (respectively, obtuse, or right) orthocentric.*

Proof. Follows immediately from Theorem 4.3.15. □

Theorem 4.3.19 *Suppose that the numbers $d_1, \dots, d_{n+1}$ are all different from zero, namely either all positive, or exactly one negative and in this case $1/\sigma = \sum_{i=1}^{n+1} 1/d_i < 0$. The $(n+1) \times (n+1)$ matrix*

$$M = \begin{bmatrix} P & Z \\ Z^T & Q \end{bmatrix}, \tag{4.33}$$

where

$$P = \begin{bmatrix} d_1 + d_2 & -d_2 & 0 & \dots & -d_1 \\ -d_2 & d_2 + d_3 & -d_3 & \dots & 0 \\ \dots & \dots & \dots & \dots & \dots \\ -d_1 & 0 & \dots & -d_{n+1} & d_1 + d_{n+1} \end{bmatrix},$$

$$Q = \begin{bmatrix} \frac{1}{d_1} - \frac{\sigma}{d_1^2} & -\frac{\sigma}{d_1 d_2} & -\frac{\sigma}{d_1 d_3} & \dots & -\frac{\sigma}{d_1 d_{n+1}} \\ -\frac{\sigma}{d_1 d_2} & \frac{1}{d_2} - \frac{\sigma}{d_2^2} & -\frac{\sigma}{d_2 d_3} & \dots & -\frac{\sigma}{d_2 d_{n+1}} \\ \dots & \dots & \dots & \dots & \dots \\ -\frac{\sigma}{d_1 d_{n+1}} & -\frac{\sigma}{d_2 d_{n+1}} & \dots & \dots & \frac{1}{d_{n+1}} - \frac{\sigma}{d_{n+1}^2} \end{bmatrix}$$

and Z is defined in (4.29), is then positive semidefinite of rank n.

Proof. Denote by I the identity matrix, by D the diagonal matrix with diagonal entries $d_1, d_2, \dots, d_{n+1}$, by C the matrix of order $n+1$

$$C = \begin{bmatrix} 0 & 1 & 0 & \dots & 0 \\ 0 & 0 & 1 & \dots & 0 \\ \dots & \dots & \dots & \dots & \dots \\ 0 & 0 & 0 & \dots & 1 \\ 1 & 0 & 0 & \dots & 0 \end{bmatrix} \quad \text{and } e = [1, 1, \dots, 1]^T.$$

Then (C^T is the transpose of C)

$$M = \begin{bmatrix} (I - C)D(I - C^T) & I - C \\ I - C^T & D^{-1} - \sigma D^{-1} e e^T D^{-1} \end{bmatrix}.$$

However

$$(I - C)D(D^{-1} - \sigma D^{-1} e e^T D^{-1}) = I - C,$$

so that

$$\begin{bmatrix} I & -(I - C)D \\ 0 & I \end{bmatrix} M \begin{bmatrix} I & 0 \\ -D(I - C^T) & I \end{bmatrix} = \begin{bmatrix} 0 & 0 \\ 0 & D^{-1} - \sigma D^{-1} e e^T D^{-1} \end{bmatrix}.$$

Since $(D^{-1} - \sigma D^{-1} e e^T D^{-1})e = D^{-1}e - D^{-1}e\sigma^{-1}(e^T D^{-1} e) = 0$, the rank of the matrix M is at most n. The formula (4.32) shows that all the principal minors of the matrix $(I - C)D(I - C^T)$ of orders $1, 2, \dots, n$ are positive. $\square$

Theorem 4.3.20 *A normal polygon $V = [A_1, A_2, \ldots, A_{n+1}]$ is acutely cyclic if and only if there exist positive numbers $p_1, p_2, \ldots, p_{n+1}$ such that the vectors $a_i = \overrightarrow{A_iA_{i+1}}$ $(i = 1, \ldots, n+1;\ A_{n+2} = A_1)$ and the number $p = \sum_{k=1}^{n+1} p_k$ satisfy*

$$\begin{aligned} \langle a_i, a_j \rangle &= \frac{1}{p} p_i p_j,\ i \neq j, \\ \langle a_i, a_i \rangle &= \frac{1}{p} p_i (p - p_i). \end{aligned} \tag{4.34}$$

This polygon is obtusely cyclic if and only if there exist numbers p_1, $p_2, \ldots, p_{n+1}$, one of which is negative, the remaining positive, and their sum $p = \sum_{i=1}^{n+1} p_i$ negative, again fulfilling (4.34).

Proof. This follows immediately from Theorems 4.3.8, 4.3.15, and 4.3.20, where d_i is set as $1/p_i$; the matrices P and Q in (4.33) are then the matrices of the orthocentric normal polygon and of the polygon from (4.34). □

Another metric characterization of cyclic normal polygons is the following:

Theorem 4.3.21 *A normal polygon $V = [A_1, A_2, \ldots, A_{n+1}]$ is acutely cyclic if and only if there exist positive numbers $p_1, p_2, \ldots, p_{n+1}$ such that the distances $\sqrt{m_{ij}} = \rho(A_i, A_j)$ $(i < j)$ satisfy*

$$m_{ij} = \frac{1}{p}(p_i + p_{i+1} + \cdots + p_{j-1})(p_j + p_{j+1} + \cdots + p_{n+1} + p_1 + \cdots + p_{i-1}), \tag{4.35}$$

where $p = \sum_1^{n+1} p_j$.

This polygon is obtusely cyclic if and only if there exist numbers p_1, $p_2, \ldots, p_{n+1}$ such that one of them is negative, the remaining positive, and their sum negative, for which again (4.35) *holds.*

Proof. Since in the notation of Theorem 4.3.20

$$m_{ij} = \langle a_i + a_{i+1} + \cdots + a_{j-1}, a_i + a_{i+1} + \cdots + a_{j-1} \rangle,$$

it suffices to show that the relations (4.34) and (4.35) are equivalent. This is done easily by induction with respect to $j - i$. □

Theorem 4.3.22 *Every m-dimensional $(m \geq 2)$ face Σ' of a cyclic n-simplex Σ is also cyclic, namely of the same kind (acutely, obtusely, or right) as that of Σ. In addition, the cyclic ordering of the vertices in Σ' is induced by that of Σ.*

Proof. Suppose $V = [A_1, A_2, \ldots, A_{n+1}]$ is a cyclic normal polygon corresponding to the simplex Σ. Let $A_{k_1}, A_{k_2}, \ldots, A_{k_{m+1}}$ $(k_1 < k_2 < \cdots < k_{m+1})$ be the vertices of the face Σ'. It suffices to show that $V' = [A_{k_1}, A_{k_2}, \ldots, A_{k_{m+1}}]$

is a cyclic polygon in the corresponding m-dimensional space $\mathcal{E}_m$. If V is a right cyclic polygon, the assertion is correct by Theorem 4.1.7. If V is an acutely or obtusely cyclic polygon, there exist by Theorem 4.3.21 numbers $p_1, p_2, \dots, p_{n+1}$ (in the first case all positive; in the second case one negative, the remaining positive, with negative sum) for which (4.35) holds. Denote

$$\sum_{k=k_1}^{k_2-1} p_k = q_1, \ \sum_{k=k_2}^{k_3-1} p_k = q_2, \ \dots, \ \sum_{k=k_m}^{k_{m+1}-1} p_k = q_m,$$

$$p_{k_{m+1}} + \dots + p_{n-1} + p_1 + \dots + p_{k_1-1} = q_{m+1}.$$

Since $p = \sum_{i=1}^{n+1} p_i = \sum_{i=1}^{m+1} q_i = q$, then either all the q_js are positive (in the first case), or exactly one of the numbers q_k is negative (namely that in whose sum the negative number p_l enters); we also have then $q < 0$.

By the formulae (4.35), it now follows that for $i < j$

$$m_{k_i k_j} = \frac{1}{q}(q_i + q_{i-1} + \dots + q_{j-1})(q_j + \dots + q_{m+1} + q_1 + \dots + q_{i-1}).$$

Indeed, V' and Σ' are cyclic of the same kind as Σ. □

Before formulating the main theorem about cyclic normal polygons, we recall the following lemma (the proof is in [11]). There we call a solution $x_0, x_1, \dots, x_m$ of the system

$$\begin{aligned} x_1(x_1 + x_2 + \dots + x_m) &= x_0 c_1, \\ x_2(x_1 + x_2 + \dots + x_m) &= x_0 c_2, \\ &\dots \\ x_m(x_1 + x_2 + \dots + x_{m-1}) &= x_0 c_m, \\ x_0^2 &= 1, \end{aligned} \tag{4.36}$$

feasible, if either all the numbers $x_1, x_2, \dots, x_m$ are positive (and then $x_0 = 1$), or exactly one of the numbers $x_1, x_2, \dots, x_m$ is negative, the remaining positive with $\sum_{i=1}^{m} x_i$ negative (and then $x_0 = -1$).

Theorem 4.3.23 *Suppose $c_1, c_2, \dots, c_m$ $(m \geq 3)$ are positive numbers, and $x_0, x_1, \dots, x_m$ is a feasible solution of the system (4.36). Then the following holds:*

(i) *If for some index k, $1 \leq k \leq m$*

$$\sqrt{c_k} \geq \sum_{j=1 j\neq i}^{m} \sqrt{c_j} \qquad \textit{or} \qquad c_k = \sum_{j=1, j\neq i}^{m} c_j,$$

then no feasible solution of the system (4.36) exists.

(ii) *If for every index $k = 1, 2, \ldots, m$ the inequality*

$$\sqrt{c_k} < \sum_{j=1, j\neq i}^{m} \sqrt{c_j}$$

is fulfilled, together with

$$c_k \neq \sum_{j=1, j\neq i}^{m} c_j,$$

then there exists one feasible solution of the system (4.36); this solution is positive if for every index k, $1 \leq k \leq m$

$$c_k < \sum_{j=1, j\neq k}^{m} c_j,$$

and nonpositive (with $x_0 = -1$), if for some k

$$c_k > \sum_{j=1, j\neq k}^{m} c_j.$$

In this last case $x_k < 0$.

We are now able to prove the main theorem.

Theorem 4.3.24 *A cyclic normal polygon in $\mathcal{E}_n$ is uniquely determined by the lengths of its $n+1$ edges and their ordering, apart from its position in the space. If $l_1, l_2, \ldots, l_{n+1}$ are these lengths, and, say, $l_{n+1} = \max(l_1, \ldots, l_{n+1})$, then there exists such a cyclic normal polygon if and only if*

$$l_{n+1} < \sum_{i=1}^{n} l_i. \tag{4.37}$$

This cyclic polygon is then acutely right, or obtusely cyclic, according to whether

$$\begin{aligned} l_{n+1}^2 &< \sum_{i=1}^{n} l_i^2, \text{ or} \\ l_{n+1}^2 &= \sum_{i=1}^{n} l_i^2, \text{ or} \\ l_{n+1}^2 &> \sum_{i=1}^{n} l_i^2. \end{aligned} \tag{4.38}$$

Proof. The necessity of condition (4.37) is geometrically clear. The right cyclic polygons fulfill the second relation in (4.38) by Theorem 4.32 and such a polygon is uniquely determined by the lengths l_i.

The second relation in (4.34) implies that for every acutely (respectively, obtusely) cyclic polygon there exists a positive (respectively, nonpositive) feasible solution of the system (4.36) for $m = n+1$, $c_i = l_i^2$, and

$$x_i = \frac{p_i}{\sqrt{|\sum p_k|}}, \qquad x_0 = \operatorname{sgn} \sum p_k, \qquad i = 1, \ldots, n+1.$$

The same relation shows that for every positive (respectively, nonpositive) feasible solution of the system (4.36) and $c_i = l_i^2$ there exists an acutely (respectively, obtusely) cyclic polygon with given lengths of edges (by setting $p_i = x_i \left| \sum_{k=1}^{n+1} x_k \right|$ in (4.34)). Thus the existence and uniqueness follow from Theorem 4.3.23 and (4.37). □

Remark 4.3.25 The relations (4.37) and (4.38) generalize the triangle inequality and the Pythagorean conditions in the case $n = 2$.

Definition 4.3.26 For a moment, we call two oriented normal polygons in $\mathcal{E}_n$ *directly* (respectively, *indirectly*) *equivalent*, if there exists such a bijection between the vectors of their edges in which the corresponding vectors are equal (respectively, opposite).

Remark 4.3.27 It is clear that every normal polygon V_2 directly equivalent to a normal polygon V_1 is obtained by permuting the vectors of oriented edges and putting them one after another, possibly changing the position by translation.

It also can be shown that all perpendicularly inscribed polygons (cf. Definition 4.3.26) into a simplex are equivalent.

Theorem 4.3.28 *All polygons, directly or indirectly equivalent to a cyclic polygon, are again cyclic, even of the same kind, and their circumscribed hyperspheres have the same radii. All (cyclic) polygons perpendicularly inscribed into an orthocentric simplex have the circumscribed hypersphere in common. The center of this hypersphere is the Lemoine point of the simplex.*

Proof. The first part follows from Theorems 4.3.8 and 4.3.19. The second part is, in the case of a right simplex, a consequence of the fact that the altitude from the vertex (which is at the same time the orthocenter) is the longest edge of all right cyclic polygons perpendicularly inscribed into the simplex. To complete the proof of the second part for the acutely or obtusely cyclic polygons, we use the relations (4.35) for the squares m_{ik} of the lengths of its edges. A simple check shows that the numbers q_{rs} (cf. Chapter 1) are given by $(i, j = 1, \ldots, n+1)$

$$q_{00} = \frac{1}{p^2}\left[\sum_{i,j=1, i\neq j}^{n+1} p_i^2 p_j + \sum_{i,j,k=1, i<j<k}^{n-1} p_i\, p_j\, p_k\right], \; q_{0i} = -\frac{1}{p}(p_{i-1} + p_i),$$

$$q_{ii} = \frac{1}{p_{i-1}} + \frac{1}{p_i}, \qquad q_{i-1,i} = q_{i,i-1} = \frac{1}{p_{i-1}}, \quad p = \sum_{k=1}^{n+1} p_k,$$
$$q_{ij} = 0 \quad \text{for } j \not\equiv i-1,\ j \not\equiv i,\ j \not\equiv i+1 \mod (n+1), \quad p_0 = p_{n+1}.$$

By Theorem 2.1.1, the radius of the circumscribed hypersphere of the simplex is $\frac{1}{2}\sqrt{q_{00}}$. This number is, however, a symmetric expression in the values of $p_1, p_2, \ldots, p_{n+1}$. Consequently, the radii of the circumscribed hyperspheres of every polygon equivalent to the given are all equal.

The last assertion follows from the second and from Theorem 4.3.8. □

The relations in the previous proof for the entries of the Gramian of a cyclic n-simplex yield, together with the three possibilities for the numbers p_i (all positive for the acute simplex; one negative, the remaining positive and the sum negative for the obtuse simplex, and the Schlaefli simplex), a complete characterization of the possibilities for extended signed graphs of cyclic simplexes.

Theorem 4.3.29 *The signed extended graph of an acute cyclic n-simplex is a wheel (cf. Appendix, Section 6.2) with $n+2$ nodes. The signed extended graph of an obtuse cyclic n-simplex is a positive circuit with $n+2$ nodes in which one node u is connected with all the not-neighboring nodes by negative edges (diagonals) and the two neighboring nodes to u are joined by a negative edge. The extended signed graph of the right cyclic simplex is, of course, a positive circuit.*

In the conclusion, we mention nets on a simplex.

Definition 4.3.30 Let an n-simplex Σ be given. By a *net* on Σ, we understand the set $N(\Sigma)$ consisting of all the vertices of Σ and a subset of edges of Σ. If, in addition, a length to every edge in $N(\Sigma)$ is assigned, we speak about a *metric net* on Σ.

Remark 4.3.31 A net on an n-simplex can be considered as a mapping of an undirected graph with $n+1$ nodes on the zero- and one-dimensional structure of Σ. We can thus speak about connectivity of a net, of nets which are trees, etc.

Suppose now that Σ is some n-simplex and N is a net on Σ. We say that an n-simplex Σ' is *N-equivalent* to Σ if there is a bijection between the set of vertices of Σ and Σ' such that the lengths of edges in N and the corresponding edges in Σ' are equal.

We can now formulate and prove a result which was established in [16].

Theorem 4.3.32 *Let N be a net on the n-simplex Σ. Then there exists an n-simplex Σ' which is N-equivalent to Σ having the maximal n-dimensional volume if and only if the net N is connected. The simplex Σ' is then* *characterized by the fact that each interior angle opposite an edge not contained in N is right.*

Proof. Geometrically evident is the fact that if N is not connected, then some distances in Σ' can be arbitrarily large and the volume is not bounded from above. However, if N is connected and one vertex of Σ' is fixed, then all possible simplexes are within some hypersphere and the volume is bounded from above. Consider the volume of all such simplexes, including those "degenerate" ones having n-dimensional volume zero.

By (1.28), the formula for the volume V using the lengths of edges is

$$V^2 = \frac{(-1)^{n+1}}{2^n (n!)^2} \det[m_{rs}].$$

Since this function is a differentiable function of the variables corresponding to edges of Σ not contained in N, on a compact set, it attains its maximum, and in the attained solution, $\partial \det[m_{rs}]/\partial m_{uv} = 0$, whenever $u \neq v$ and $A_u A_v \notin N$. This partial derivative is proportional to the (u, v) entry of the inverse matrix of $[m_{rs}]$ which is (cf. formula (1.21)) $-2q_{uv}$ so that $q_{uv} = 0$. This means that the interior angles corresponding to all edges not contained in N are right as asserted. □

Here are two important corollaries.

Theorem 4.3.33 *If N consists of the edges A_1A_2, A_2A_3, $\ldots$, A_nA_{n-1}, $A_{n-1}A_1$, the maximum n-simplex is cyclic. The simplex is by the metric net determined uniquely up to the position in the space.*

Proof. By Theorem 4.3.32, the maximal simplex is cyclic. The rest follows from Theorem 4.3.24. □

Theorem 4.3.34 *If the metric net is a tree, the corresponding simplex is right and the given edges are its legs.*

Proof. This follows again from Theorem 4.3.32 and the theory of right simplexes. □

Remark 4.3.35 Observe that Theorem 4.3.32 confirms the conjecture 2.1.9 from Chapter 2 in the case that all the given angles are right.

4.4 Simplexes with a principal point

In this short section, we present a class of special simplexes which contains generalizations of some properties of the triangle. It is defined by the corresponding Menger matrix as follows:

We say that the set of points $A_1, \ldots, A_{n+1}$ in $\mathcal{E}_n$ forms vertices of an *n-simplex with a principal point* if the points are linearly independent and the squares of distances between A_i and A_j, $i \neq j$, satisfy

$$|A_iA_j|^2 = \alpha(t_i^2 + t_j^2) + \beta t_i t_j \tag{4.39}$$

for some real $t_1, \ldots, t_{n+1}$, α and β, whenever $i \neq j$ and $i, j = 1, \ldots, n+1$.

Remark 4.4.1 We proved in [5] that necessary and sufficient conditions for the existence of an n-simplex satisfying (4.39) are that

$$\alpha + \beta > 0$$

and one of the following holds:

(i) all the t_is are different from zero and

$$\beta\left(\sum_{i=1}^{n+1} \frac{1}{t_i}\right)^2 + (\alpha - n\beta)\sum_{i=1}^{n+1} \frac{1}{t_i^2} > 0; \tag{4.40}$$

(ii) one of the t_is is zero and $\alpha > (n-1)\beta$.

As is well known, if a triangle has all angles less than $\frac{2}{3}\pi$, there exists in the triangle a point from which all the sides are seen at the angle $\frac{2}{3}\pi$. This point is called the *Toricelli point.* The following theorem is a generalization.

Theorem 4.4.2 *A necessary and sufficient condition that an n-simplex with vertices $A_1, \ldots, A_{n+1}$ possesses a point $P \neq A_i$ (for all i) such that the angles $\angle A_iPA_j$ are all equal, is that* (4.39) *holds for $\alpha = n\beta$ and all the t_is have the same sign.*

Proof. Suppose that such a point exists. Then the points on the halflines PA_i having the same positive distance from P form vertices of a regular n-simplex. As we proved in Theorem 4.39, the common angle $\psi = \angle A_iPA_j$ satisfies $\cos\psi = -1/n$. If we denote the distance PA_i as t_i, we obtain by the cosine theorem

$$|A_iA_j|^2 = t_i^2 + t_j^2 + \frac{2}{n}t_i t_j,$$

i.e. (4.39) for $\alpha = n\beta$ (up to a positive multiple).

The converse is also easily established. □

Another property of the triangle is the following. If P_1, P_2, and P_3 are the points in which the incircle touches the sides of the triangle $A_1A_2A_3$, then the segments A_iP_i meet at one point, called the *Gergonne point* of the triangle. For a simplex, we have two generalizations.

Theorem 4.4.3 *A necessary and sufficient condition that in an n-simplex there exists a hypersphere which touches all edges (as segments) is that for* *the squares of the lengths of edges,* (4.39) *holds for* $\alpha = \beta$ *positive, and all the* t_i*s have the same sign. In addition, all the hyperplanes determined by any* $n-1$ *vertices and that point on the opposite edge in which the hypersphere touches the edge meet at one point.*

Proof. Suppose that such a hypersphere H exists. If A_i is a vertex, then all the points P_{ij} in which H touches the edges A_iA_j have the same distance t_i from A_i. Since $|A_iA_j|^2 = t_i^2 + t_j^2 + 2t_it_j$, (4.39) holds for $\alpha = \beta$ and $t_i > 0$. The converse also holds. The point with barycentric coordinates $(1/t_i)$ has then the mentioned property. □

The second possibility is the following:

Theorem 4.4.4 *A necessary and sufficient condition that for the inscribed hypersphere the points of contacts* P_i *in the* $(n-1)$*-dimensional faces have the property that the lines* A_iP_i *meet at one point is that for the squares of the lengths of edges,* (4.39) *holds for* $\alpha = (n-1)\beta$ *positive, and the* t_i*s of the same sign.*

Proof. First, we shall show that for an nb-point $Q = (q_1, \ldots, q_{n+1})$ there is just one quadric which touches all the $(n-1)$-dimensional faces in the projection points of Q on these faces, namely the quadric with equation

$$\sum_i \left(\frac{x_i}{q_i}\right)^2 - \left(\sum_i \frac{x_i}{q_i}\right)^2 = 0. \tag{4.41}$$

Indeed, let

$$\sum_{i,k} c_{ik}\xi_i\xi_k = 0$$

be the dual equation of such a quadric. Then $c_{ii} = 0$ for all i since the face ω_i is the tangent dual hyperplane, and thus

$$\sum_k c_{ik}\xi_k = 0$$

is the equation of the tangent point. Therefore, $c_{ik} = \lambda_i q_k$ for some nonzero constant λ_i. Since $c_{ik} = c_{ki}$ for all i, k, we obtain altogether $c_{ik} = \lambda q_i q_k$ for all i, k, $i \neq k$. The matrix of the dual quadric is thus a multiple of

$$Z = Q_0(J - I)Q_0,$$

where Q_0 is the diagonal matrix diag $(q_1, \ldots, q_{n+1})$, J is the matrix of all ones, and I the identity matrix. The inverse of Z is thus a multiple of $Q_0^{-1}(nI - J)Q_0^{-1}$, which exactly corresponds to (4.41).

Now, the equation (4.41) has to be the equation of a hypersphere, thus of the form

$$\alpha_0 \sum_{i,k} m_{ik} x_i x_k - 2 \sum_k \alpha_k x_k \sum_j x_j = 0, \ \alpha_0 \neq 0.$$

Comparing both equations, we obtain

$$\alpha_i = -\frac{n-1}{2q_i^2},$$

and for $i \neq j$

$$-2\alpha_0 m_{ij} = (n-1)\left(\frac{1}{q_i^2} + \frac{1}{q_j^2}\right) + \frac{2}{q_i q_j},$$

so that (4.41) holds for $\alpha = (n-1)\beta$ and $t_i = 1/q_i$. □

A generalization of so-called *isodynamical centers* of the triangle is the following theorem. In the formulation, if A, B, and C are distinct points, $H(A, B; C)$ will be the Apollonius hypersphere, the locus of the points X where the ratio of distances $|XA|$ and $|XB|$ is the same as that of $|CA|$ and $|CB|$.

Theorem 4.4.5 *Let $A_1, \dots, A_{n+1}$ be vertices of a simplex in $\mathcal{E}_n$. Then a necessary and sufficient condition that all the hyperspheres $H(A_i, A_j; A_k)$, for i, j, k distinct, have a proper point in common is that* (4.39) *holds for $\alpha = 0$, and $t_i \neq 0$ for all i.*

Proof. Suppose that all the hyperspheres $H(A_i, A_j; A_k)$ have a point $D = (d_1, \dots, d_{n+1})$ in common. The equation of $H(A_i, A_j; A_k)$ is (using (1.9))

$$\begin{aligned} m_{ik}&\left[\sum_{p,q} m_{pq} x_p x_q - 2 \sum_p m_{jp} x_p \sum_q x_q\right] \\ &-m_{jk}\left[\sum_{p,q} m_{pq} x_p x_q - 2 \sum_p m_{ip} x_p \sum_q x_q\right] = 0. \end{aligned} \tag{4.42}$$

Denote as t_i the numbers $\sum_{p,q} m_{pq} d_p d_q - 2 \sum_p m_{ip} d_p \sum_q d_q$; since they are proportional to the squares of distances between D and the A_is, at least one t_i, say t_l, is different from zero. By $m_{ik} t_l = m_{il} t_k$, it follows that all the t_ks are different from zero. By (4.42), all the numbers $m_{ik}/t_i t_k$ are equal, so that indeed $m_{ik} = \rho t_i t_k$ as asserted.

The converse is also easily established. □

Another type of simplex with a principal point will be obtained from the so-called $(n+1)$-*star* in $\mathcal{E}_n$. It is the set of $n+1$ halflines, any two of which span the same angle. This $(n+1)$-tuple is congruent to the $(n+1)$-tuple of halflines CA_i in a regular n-simplex with vertices A_i and centroid C. Therefore, the angle ω between any two of these halflines satisfies, as we shall see in Theorem 4.5.1, $\cos \omega = -1/n$.

Theorem 4.4.6 *A necessary and sufficient condition that an n-simplex* Σ *with vertices* $A_1, \dots, A_{n+1}$ *in* $\mathcal{E}_n$ *has the property that in a Euclidean* $(n+1)$*-dimensional space containing* $\mathcal{E}_n$ *there exists a point* A *such that the halflines* $AA_1, \dots, AA_{n+1}$ *can be completed by a halfline* AA_0 *into an* $(n+2)$*-star is that the squares of the lengths of edges of* Σ *fulfill* (4.39) *with* $\alpha = -(n+1)\beta$ *positive, and the* t_i*s have the same sign.*

Proof. Suppose such a point A exists. Since the cosine of the angle A_iAA_j is $-\frac{1}{n+1}$, the cosine theorem implies that, denoting $|AA_i|$ as t_i

$$|A_iA_j|^2 = t_i^2 + t_j^2 + \frac{2}{n+1} t_i t_j, \tag{4.43}$$

whenever $i \neq j$. The converse is also easily established. □

Recalling the property (4.7) from Theorem 4.2.1, we have also

Theorem 4.4.7 *A necessary and sufficient condition that an n-simplex is acute orthocentric is that the squares of the lengths of its edges fulfill* (4.39) *with* α *positive and* $\beta = 0$ *(and all the* t_i*s nonzero).*

Remark 4.4.8 We did not complicate the theorems by allowing some of the t_is to be negative. This case is, however, interesting, similarly as in Theorem 4.2.8 for the orthocentric $(n+2)$-tuple, in one case: if $\alpha = -(n+1)\beta$ is positive and t_0 satisfies

$$\sum_{i=0}^{n+1} \frac{1}{t_i} = 0.$$

In this case, the whole $(n+2)$-tuple behaves symmetrically and (4.43) holds for all $n+2$ points.

4.5 The regular n-simplex

For completeness, we list some well-known properties of the *regular* n-simplex, which has all edges of the same length.

Theorem 4.5.1 *If all edges of the regular n-simplex* Σ *have length one, then the radius of the circumscribed hypersphere is* $\sqrt{\frac{n}{2(n+1)}}$, *the radius of the inscribed hypersphere is* $\frac{1}{\sqrt{2n(n+1)}}$, *all the interior dihedral angles are equal to* φ, *for which* $\cos\varphi = 1/n$, *and all edges are seen from the centroid by the angle* ψ *satisfying* $\cos\psi = -1/n$. *All distinguished points, such as the centroid, the center of the circumscribed hypersphere, the center of the inscribed hypersphere, the Lemoine point, etc., coincide. The Steiner cicumscribed ellipsoid coincides with the circumscribed hypersphere.*

Proof. If e is the vector with $n+1$ coordinates, all equal to one, then the Menger matrix of Σ is

$$\begin{bmatrix} 0 & e^T \\ e & ee^T - I \end{bmatrix}.$$

Since

$$\begin{bmatrix} 0 & e^T \\ e & ee^T - I \end{bmatrix} \begin{bmatrix} \frac{2n}{n+1} & -\frac{2}{n+1}e^T \\ -\frac{2}{n+1}e & -\frac{2}{n+1}ee^T + 2I \end{bmatrix} = -2I_{n+2},$$

the values of the entries q_{rs} of the extended Gramian of Σ result. The radii then follow from the formulae in Corollary 1.4.13 and in Theorem 2.2.1, the angle from $\cos\varphi_{ik} = -q_{ik}/\sqrt{q_{ii}q_{kk}}$. The rest is obvious. □

Remark 4.5.2 It is clear that the regular n-simplex is hyperacute, even totally hyperacute, and orthocentric.

5
Further geometric objects

We begin with an involutory relationship within the class of all n-simplexes.

5.1 Inverse simplex

It is well known (e.g. [28]) that for every complex (or, real) matrix (even not necessarily square) A, there exists a unique complex matrix A^+ with the properties

$$\begin{aligned} AA^+A &= A, \\ A^+AA^+ &= A^+, \\ (AA^+)^* &= AA^+, \\ (A^+A)^* &= A^+A, \end{aligned}$$

where $*$ means conjugate transpose.

The matrix A^+ is called the *Moore–Penrose inverse* of A and clearly constitutes with A an involution in the class of all complex matrices

$$(A^+)^+ = A.$$

In addition, the following holds:

Theorem 5.1.1 *If A is a (real) symmetric positive semidefinite matrix, then A^+ is also a symmetric positive semidefinite matrix, and the sets of vectors x, for which $Ax = 0$ and $A^+x = 0$, coincide. More explicitly, if*

$$A = U \begin{bmatrix} D & 0 \\ 0 & 0 \end{bmatrix} U^T,$$

where U is orthogonal and D is nonsingular diagonal, then

$$A^+ = U \begin{bmatrix} D^{-1} & 0 \\ 0 & 0 \end{bmatrix} U^T.$$

Proof. This is a consequence of the well-known theorem ([28], p.64) on the singular value decomposition. □

We shall also use a result from [27]. Recall that in Chapter 1, Theorem 1.1.2, we saw that for n ordered linearly independent vectors $u_1, \ldots, u_n$ in a Euclidean space E_n there exists another linearly independent ordered system of n vectors $v_1, \ldots, v_n$ such that the Gram matrix of $2n$ vectors u_i, v_i has the form

$$\begin{bmatrix} G(u) & I \\ I & G(v) \end{bmatrix},$$

and this matrix has rank n. These two ordered n-tuples of vectors were called biorthogonal bases in E_n.

We use the generalization of the biorthogonal system defined in [27]:

Theorem 5.1.2 *Let $u_1, \ldots, u_m$ be a system of m vectors in E_n with rank (maximum number of linearly independent vectors of the system) n. Then there exists a unique system of m vectors $v_1, \ldots, v_m$ in E_n such that the Gram matrix of the $2m$ vectors u_i, v_i has the form*

$$G_1 = \begin{bmatrix} G(u) & P \\ P^T & G(v) \end{bmatrix},$$

where the matrix P satisfies $P^2 = P$, $P = P^T$, and this matrix G_1 has rank n. In fact, we can find the matrix P as follows: $P = I - R(R^TR)^{-1}R^T$, where R is an arbitrary matrix whose columns are formed by the maximal number of linearly independent vectors x satisfying $G(u)x = 0$; this means that these x's are the vectors of the linear dependence relations among the vectors u_i.

Remark 5.1.3 The matrix $G(v)$ is then the Moore–Penrose inverse of $G(u)$. Moreover, the vectors v_i fulfill the same linear dependence relations as the vectors u_i. We shall call the system of vectors v_i the *generalized biorthogonal system* to the system u_i.

All these properties will be very useful for the Gramian Q of an n-simplex Σ. In this case, we have a system of $n+1$ privileged vectors in E_n, namely the system of (normalized in a sense) outer normals whose Gram matrix is the Gramian Q.

Theorem 5.1.4 *For the n-simplex Σ with vertices $A_1, \ldots, A_{n+1}$ and centroid C, there exists a unique n-simplex Σ^{-1} such that for its vertices $B_1, \ldots, B_{n+1}$ the following holds: the vectors $\overrightarrow{CB_i}$ form the generalized biorthogonal system to the vectors $\overrightarrow{CA_i}$.*

We shall call this simplex Σ^{-1} the *inverse simplex* of the simplex Σ.

Remark 5.1.5 The inverse simplex to the inverse simplex is clearly the original simplex.

Let us describe now the metric relations between the two simplexes using their characteristics in the Menger matrices and Gramians.

The vectors $u_i = \overrightarrow{CA_i}$ for $i = 1, \ldots, n+1$, are the vectors of the medians and they satisfy a single (linearly independent) relation $\sum_i u_i = 0$. Using the result in Theorem 5.1.2, we obtain

$$P = I - \frac{1}{n+1} J, \tag{5.1}$$

where $J = ee^T$, $e = [1, \ldots, 1]^T$ so that $\langle u_i, v_j \rangle = \frac{n}{n+1}$ for $i = j$, and $\langle u_i, v_j \rangle = -\frac{1}{n+1}$ for $i \neq j$. Thus if i, j, k are distinct indices, then $\langle u_i, v_j - v_k \rangle = 0$, which means that the vector u_i is orthogonal to the $(n-1)$-dimensional face $\hat{\omega}_i$ of the simplex Σ^{-1} (opposite to the vertex B_i). It is thus the vector of the (as can be shown, outer) normal of Σ^{-1}.

Let us summarize that in a theorem:

Theorem 5.1.6 *The vectors of the medians of a simplex are the outer normals of the inverse simplex. Also, the medians of the inverse simplex are the outer normals of the original simplex.*

Remark 5.1.7 This statement does not specify the magnitudes of the simplexes. In fact, the unit in the space plays a role.

Let us return to the metric facts. Here, P will again be the matrix from (5.1). We start with a lemma.

Lemma 5.1.8 *Let $\mathcal{X}$ be the set of all $m \times m$ real symmetric matrices $X = [x_{ij}]$ satisfying $x_{ii} = 0$ for all i, and let $\mathcal{Y}$ be the set of all $m \times m$ real symmetric matrices $Y = [y_{ij}]$ satisfying $Ye = 0$. If P is the $m \times m$ matrix $P = I - \frac{1}{m} ee^T$ as in (5.1), then the following are equivalent for two matrices X and Y:*

(i) $X \in \mathcal{X}$ *and* $Y = -\frac{1}{2} PXP \in \mathcal{Y}$;
(ii) $Y \in \mathcal{Y}$ *and* $x_{ik} = y_{ii} + y_{kk} - 2y_{ik}$ *for all* i, k;
(iii) $Y \in \mathcal{Y}$ *and* $X = ye^T + ey^T - 2Y$, *where* $y = [y_{11}, \ldots, y_{nn}]^T$.

In addition, if these conditions are fulfilled, then X is a Menger matrix (for the corresponding m) if and only if the matrix Y is positive semidefinite of rank $m-1$.

Proof. The conditions (ii) and (iii) are clearly equivalent. Now suppose that (i) holds. Then $Y \in \mathcal{Y}$. Define the vector $y = \frac{1}{n} Xe - (\text{tr} Y)e$, where $\text{tr}\, Y$ is the trace of the matrix Y, $\sum_i y_{ii}$. Then

$$ye^T + ey^T - 2Y = X;$$

since $x_{ii} = 0$, we have $y = [y_{11}, \ldots, y_{nn}]^T$, i.e. (iii). Conversely, assume that (iii) is true. Then $x_{ii} = 0$ for all i, so that $X \in \mathcal{X}$, and also $-\frac{1}{2}PXP = Y$, i.e. (i) holds.

To complete the proof, suppose that (i), (ii), and (iii) are fulfilled and that X is a Menger matrix, so that $X \in \mathcal{X}$. If z is an arbitrary vector, then $z^TYz = -\frac{1}{2}z^TPXPz$, which is nonnegative by Theorem 1.2.4 since $u = Pz$ fulfills $e^Tu = 0$.

Suppose conversely that $Y \in \mathcal{Y}$ is positive semidefinite. To show that the corresponding matrix X is a Menger matrix, let u satisfy $e^Tu = 0$. Then $u^TXu = u^T(ye^T + ey^T - 2Y)u$, which is $-2u^TYu$, and thus nonpositive. By Theorem 1.2.4, X is a Menger matrix. □

For simplexes, we have the following result:

Theorem 5.1.9 *Suppose that $\mathcal{A}$, $\mathcal{B}$ are the (ordered) sets of vertices of two mutually inverse n-simplexes Σ and Σ^{-1}. Let $P = I - \frac{1}{n+1}J$. Then the Menger matrices M_Σ and $M_{\Sigma^{-1}}$ satisfy the condition: the matrices $\frac{1}{2}PM_\Sigma P$ and $\frac{1}{2}PM_{\Sigma^{-1}}P$ are mutual Moore–Penrose inverse matrices.*

Also, the Gramians $Q(\Sigma)$ and $Q(\Sigma^{-1})$ of both n-simplexes are mutual Moore–Penrose inverse matrices, and

$$-\frac{1}{2}PM_{\Sigma^{-1}}P = Q(\Sigma),\ -\frac{1}{2}PM_\Sigma P = Q(\Sigma^{-1}).$$

The following relations hold between the entries m_{ik} of the Menger matrix of the n-simplex Σ and the entries $\tilde{q}_{ik}$ of the Gramian of the inverse simplex Σ^{-1}

$$m_{ik} = \tilde{q}_{ii} + \tilde{q}_{kk} - 2\tilde{q}_{ik}, \tag{5.2}$$

$$\tilde{q}_{ik} = -\frac{1}{2}\left(m_{ik} - \frac{1}{n+1}\sum_j m_{ij} - \frac{1}{n+1}\sum_j m_{kj} + \frac{1}{(n+1)^2}\sum_{j,l} m_{jl}\right). \tag{5.3}$$

Analogous relations hold between the entries of the Menger matrix of the simplex Σ^{-1} and the entries of the Gramian of Σ.

Proof. Denote by u_i the vectors $A_i - C$, where C is the centroid of the simplex Σ, and analogously let $v_i = B_i - C$. The (i,k) entry m_{ik} of the matrix M_Σ is $\langle A_i - A_k, A_i - A_k\rangle$, which is $\langle A_i - C, A_i - C\rangle + \langle A_k - C, A_k - C\rangle - 2\langle A_i - C, A_k - C\rangle$. If now the c_{ik}s are the entries of the Gram matrix $G(u)$, then $m_{ik} = c_{ii} + c_{kk} - 2c_{ik}$. By (ii) of Lemma 5.1.8, $G(u) = -\frac{1}{2}PM_\Sigma P$. Analogously, $G(v) = -\frac{1}{2}PM_{\Sigma^{-1}}P$, which implies the first assertion of the theorem.

The second part is a direct consequence of Theorem 5.5.4, since both $G(v)$ and $Q(\Sigma)$ are the Moore–Penrose inverse of the matrix $-\frac{1}{2}PM_\Sigma P$. The remaining formulae follow from Lemma 5.1.8. □

In the conclusion of this section, we shall extend the definition of the inverse simplex from Theorem 5.1.6 to the case that the number of points of the system is greater by more than one than the dimension of the system.

Definition 5.1.10 Let $\mathcal{A} = (A_1, \ldots, A_m)$ be an ordered m-tuple of points in the Euclidean point space $\mathcal{E}_n$, and let C be the centroid of the system. If $V = (v_1, \ldots, v_m)$ is the generalized biorthogonal system of the system $U = (A_1 - C, \ldots, A_m - C)$, then we call the ordered system $\mathcal{B} = (B_1, \ldots, B_m)$, where B_i is defined by $v_i = B_i - C$, the *inverse point system* of the system $\mathcal{A}$.

The following theorem is immediate.

Theorem 5.1.11 *The points of the inverse system fulfill the same linear dependence relations as the points of the original system. Also, the inverse system of the inverse system is the original system.*

Example 5.1.12 Let $A_1, \ldots, A_m$, $m \geq 3$, be points on a (Euclidean) line E_1, with coordinates $a_1, \ldots, a_m$, at least two of which are distinct. If e is the unit vector of E_1, then the centroid of the points is the point C with coordinate $c = \frac{1}{m}\sum_i a_i$ and the vectors of the "medians" are $(a_1 - c)e, (a_2 - c)e, \ldots, (a_m - c)e$. The Gram matrix G is thus the $m \times m$ matrix $[g_{ij}]$, where $g_{ij} = (a_i - c)(a_j - c)$. It has rank one and it is easily checked that the matrix G_1 from Theorem 5.1.2 is

$$G_1 = \begin{bmatrix} G & \omega G \\ \omega G & \omega^2 G \end{bmatrix},$$

where ω is such that the matrix ωG satisfies $(\omega G)^2 = \omega G$; thus ω is easily seen to be $[\sum_i (a_i - c)^2]^{-1}$. This means that the inverse set is obtained from the original set by extending it (or, diminishing it) proportionally from the centroid by the factor ω.

Let us perform this procedure in usual cartesian coordinates. Denote by Y the $m \times n$ matrix $[a_{ip}]$, $i = 1, \ldots, m$, $p = 1, \ldots, n$; the ith row is formed by the cartesian coordinates of the vector $u_i = A_i - C$ in E_n. Since $\Sigma u_i = 0$

$$e^T Y = 0.$$

Let K be the matrix $K = \frac{1}{m} Y^T Y$.[1]

Then form the system of hyperquadrics

$$x^T K^{-1} x - \rho = 0,$$

where x is the column vector of cartesian variable coordinates and ρ is a positive parameter. Such a hyperquadric (since K is positive semidefinite,

[1] The matrix K occurs in multivariate factor analysis when the points A_i correspond to measurements; it is usually called the *covariance matrix*.

it is an ellipsoid, maybe degenerate) can be brought to the simplest form, so-called principal axes form, by transforming K to the diagonal form using (vii) of Theorem A.1.37. The eigenvectors of K correspond to the vectors of the principal axes; the eigenvalues are proportional to the lengths of the halfaxes of the ellipsoid.

The eigenvalues of the matrix $K = \frac{1}{m}Y^TY$ are at the same time also the *nonzero* eigenvalues of the matrix $\frac{1}{m}YY^T$, i.e. of the matrix $\frac{1}{m}G(u)$, where $G(u)$ is the Gram matrix of the system u_i.

If thus $B_1, \ldots, B_m$ form the generalized point biorthogonal system to the system $A_1, \ldots, A_m$, and $\hat{K}$ is a similarly formed covariance matrix of the system B_i, then the eigenvalues of the matrix $\hat{K}$ are the nonzero eigenvalues of $\frac{1}{m}(G(u))^+$. For positive definite matrices, the Moore–Penrose inverses have mutually reciprocal nonzero eigenvalues and common eigenvectors. For our case, it means that the systems of elliptic quadrics of the systems A_i and B_i have the same systems of principal axes and the lengths of the halfaxes are (up to constant nonzero multiples) reciprocal. We shall show that if $m = n+1$, the Steiner ellipsoids belong to the system of the just-mentioned elliptic hyperquadrics. Indeed, the matrix $U = YK^{-1}Y^T$ has then rank n, and satisfies $U = U^T$, $U^2 = U$, and also $Ue = 0$. Therefore, $U = I - \frac{1}{n+1}J$, which means that the diagonal entries of U are mutually equal to $\frac{n}{n+1}$. It follows that the hyperquadric of the system corresponding to $\rho = \frac{n}{n+1}$ contains all the points A_i, and is thus the Steiner circumscribed ellipsoid.

5.2 Simplicial cones

Let us start with a theorem which completes the theory of biorthogonal bases (Appendix, Theorem A.1.47)

Theorem 5.2.1 *Let $a_1, \ldots, a_n$, $b_1, \ldots, b_n$, $n \geq 2$ be biorthogonal bases in E_n, $n \geq 2$, and $1 \leq k \leq n-1$. Then the biorthogonal completion of $a_1, \ldots, a_k$ in the corresponding E_k is $\hat{b}_1, \ldots, \hat{b}_k$, where, for each j, $1 \leq j \leq k$*

$$\hat{b}_j = b_j - [\langle b_j, b_{k+1}\rangle, \langle b_j, b_{k+2}\rangle, \ldots, \langle b_j, b_n\rangle]G^{-1}[b_{k+1}, \ldots, b_n]^T, \tag{5.4}$$

where G is the Gram matrix $G(b_{k+1}, \ldots, b_n)$; in other words, $\hat{b}_j$ is the orthogonal projection of b_j on E_k along the linear space spanned by $b_{k+1}, \ldots, b_n$.

Proof. It is clear that each vector $\hat{b}_j$ is orthogonal to every vector a_i for $i \neq j$, $i = 1, \ldots, k$. Also, $\langle a_j, \hat{b}_j\rangle = 1$. It remains to prove that $\hat{b}_j \in E_k$ for all $j = 1, \ldots, k$. It is clear that E_k is the set of all vectors orthogonal to all vectors $b_{k+1}, \ldots, b_n$. If now i satisfies $k+1 \leq i \leq n$, then denote $G^{-1}[\langle b_{k+1}, b_i\rangle, \ldots, \langle b_n, b_i\rangle]^T = v$, or equivalently $Gv = [\langle b_{k+1}, b_i\rangle, \ldots, \langle b_n, b_i\rangle]^T$. Thus v is the vector with all zero entries except the one with index i, equal to one. It follows by (5.4) that $\hat{b}_j$ is orthogonal to all the b_is so that $\hat{b}_j \in E_k$ and it is the orthogonal projection on E_k. □

Now let $a_1, \ldots, a_m$ be linearly independent vectors in E_n. The *cone generated* by these vectors is the set of all *nonnegative* linear combinations $\sum_{i=1}^m \lambda_i a_i$, $\lambda_i \geq 0$ for all i. We then say that a set C in E_n is a *simplicial cone* if there exist n linearly independent vectors which generate C. To emphasize the dimension, we speak sometimes about a simplicial n-cone.

Theorem 5.2.2 *Any simplicial cone C has the following properties:*

(i) *The elements of C are vectors.*
(ii) *C is a convex set, i.e. if $u \in C$, $v \in C$, then $\alpha u + \beta v \in C$, whenever α and β are nonnegative numbers satisfying $\alpha + \beta = 1$.*
(iii) *C is a nonnegatively homogeneous set, i.e. if $u \in C$, then $\lambda u \in C$, whenever λ is a nonnegative number.*

Proof. Follows from the definition. □

We can show that the zero vector is distinguished (the *vertex* of the cone) and each *vertex halfline*, i.e. the halfline, or, ray generated by one of the vectors a_i, is distinguished. Therefore, the cone C is uniquely determined by *unit generating vectors* $\tilde{a}_i = a_i/\sqrt{\langle a_i, a_i\rangle}$. If these vectors are ordered, the numbers $\gamma_1, \ldots, \gamma_n$ uniquely assigned to any vector v of E_n in $v = \gamma_1\tilde{a}_1 + \ldots + \gamma_n\tilde{a}_n$ will be called *spherical coordinates* of the vector v. Analogous to the simplex case, the $(n-1)$-dimensional *faces* $\omega_1, \ldots, \omega_n$ of the cone can be defined; the face ω_i is either the cone generated by all the vectors a_j for $j \neq i$, or the corresponding linear space. Then, these hyperplanes will be called *boundary hyperplanes*.

There is another way of determining a simplicial n-cone. We start with n linearly independent vectors, say $b_1, \ldots, b_n$ and define C as the set of all vectors x satisfying $\langle b_i, x\rangle \geq 0$ for all i. We can however show the following:

Theorem 5.2.3 *Both definitions of a simplicial n-cone are equivalent.*

Proof. In the first definition, let $c_1, \ldots, c_n$ be vectors which complete the a_is to biorthogonal bases. If a vector in the cone has the form $x = \sum_{i=1}^n \lambda_i a_i$, then for every i, $\langle c_i, x\rangle = \lambda_i$, and is thus nonnegative. Conversely, if a vector y satisfies $\langle c_i, y\rangle \geq 0$, then y has the form $\sum_{i=1}^n \langle c_i, y\rangle a_i$.

If, in the second definition, $d_1, \ldots, d_n$ are vectors completing the b_is to biorthogonal bases, then we can similarly show that the vectors of the form $\sum_{i=1}^n \lambda_i d_i$, $\lambda_i \geq 0$, completely characterize the vectors satisfying $\langle b_i, x\rangle \geq 0$ for all i. □

Remark 5.2.4 In the second definition, the expression $\langle b_i, x\rangle = 0$ represents the equation of a hyperplane ω_i in E_n and $\langle b_i, x\rangle \geq 0$ the halfspace in E_n

with boundary in that hyperplane. We can thus say that the simplicial n-cone C is the intersection of n halfspaces, the boundary hyperplanes of which are linearly independent. In fact, if we consider the vector a_i in the first definition as an analogy of the vertices of a simplex, the boundary hyperplanes just mentioned form an analogy of the $(n-1)$-dimensional faces of the n-simplex.

We thus see that to the simplicial n-cone C generated by the vectors a_i, we can find the cone $\tilde{C}$ generated by the normals b_i to the $(n-1)$-dimensional faces. By the properties of biorthogonality, the repeated construction leads back to the original cone. We call the second cone the *polar cone* to the first.

Here, an important remark is in order.

Remark 5.2.5 To every simplicial n-cone in E_n generated by the vectors $a_1, \ldots, a_n$, there exist further $2^n - 1$ simplicial n-cones, each of which is generated by the vectors $\epsilon_1 a_1, \epsilon_2 a_2, \ldots, \epsilon_n a_n$, where the epsilons form a system of ones and minus ones. These will be called *conjugate n-cones* to C.

The following is easy to prove:

Theorem 5.2.6 *The polar cones of conjugate cones of C are conjugates of the polar cone of C.*

We now intend to study the metric properties of simplicial n-cones. First, the following is important:

Theorem 5.2.7 *Two simplicial n-cones generated by vectors $a_i, \ldots, a_n$ and $\hat{a}_i, \ldots, \hat{a}_n$ are congruent (in the sense of Euclidean geometry, i.e. there exists an orthogonal mapping which maps one into the other) if and only if the matrices (of the cosines of their angles)*

$$\left[\frac{\langle a_i, a_j\rangle}{\sqrt{\langle a_i, a_i\rangle\langle a_j, a_j\rangle}}\right] \tag{5.5}$$

and

$$\left[\frac{\langle \hat{a}_i, \hat{a}_j\rangle}{\sqrt{\langle \hat{a}_i, \hat{a}_i\rangle\langle \hat{a}_j, \hat{a}_j\rangle}}\right]$$

differ only by permutation of rows and columns.

Proof. It is obvious that if the second cone is obtained by an orthogonal mapping from the first, then the angles between the mapped vertex halflines coincide so that the matrices are equal, or permutation equivalent, if renumbering of the vertex halflines is necessary.

To prove the converse, suppose that the matrices are equal, perhaps after renumbering. Then the Gram matrices $G = [\langle a_i, a_j\rangle]$ and $\hat{G} = [\langle \hat{a}_i, \hat{a}_j\rangle]$ satisfy $\hat{G} = DGD$ for some diagonal matrix D with positive diagonal entries $d_1, \ldots, d_n$. Define a mapping $\mathcal{U}$ which assigns to a vector x of the form

$x = \sum_{i=1}^n \lambda_i a_i$ the vector $\mathcal{U}x = \sum_{i=1}^n \lambda_i d_i^{-1} \hat{a}_i$. Since for any two vectors x of this form and $y = \sum_{i=1}^n \mu_i d_i^{-1} \hat{a}_i$

$$\begin{aligned} \langle \mathcal{U}x, \mathcal{U}y \rangle &= \left\langle \sum_{i=1}^n \lambda_i d_i^{-1} \hat{a}_i, \sum_{j=1}^n \mu_j d_j^{-1} \hat{a}_j \right\rangle \\ &= \sum_{i,j} \lambda_i \mu_j d_i^{-1} d_j^{-1} \langle \hat{a}_i, \hat{a}_j \rangle = \sum_{i,j} \lambda_i \mu_j \langle a_i, a_j \rangle \\ &= \left\langle \sum_{i=1}^n \lambda_i a_i, \sum_{j=1}^n \mu_j a_j \right\rangle = \langle x, y \rangle, \end{aligned}$$

the mapping $\mathcal{U}$ is orthogonal and maps the first cone onto the second. □

It follows that the matrix (5.5) determines the geometry of the cone. We shall call it the *normalized Gramian* of the n-cone.

Theorem 5.2.8 *The normalized Gramian of the polar cone is equal to the inverse of the normalized Gramian of the given simplex, multiplied by a diagonal matrix D with positive diagonal entries from both sides in such a way that the resulting matrix has ones along the diagonal.*

Proof. This follows from the fact that the vectors a_i and b_i form – up to possible multiplicative factors – a biorthogonal system so that Theorem A.1.45 can be applied. □

This can also be more explicitly formulated as follows:

Theorem 5.2.9 *If $G(C)$ and $G(\tilde{C})$ are the normalized Gramians of the cone C and the polar cone $\tilde{C}$, respectively, then there exists a diagonal matrix D with positive diagonal entries such that*

$$G(\tilde{C}) = D[G(C)]^{-1} D.$$

In other words, the matrix

$$\begin{bmatrix} G(C) & D \\ D & G(\tilde{G}) \end{bmatrix} \tag{5.6}$$

has rank n. The matrix D has all diagonal entries smaller than or equal to one; all these entries are equal to one if and only if the generators of C are totally orthogonal, *i.e. if any two of them are orthogonal.*

Proof. The matrix (5.6) is the Gram matrix of the biorthogonal system normalized in such a way that all vectors are unit vectors. Since the ith diagonal entry d_i of D is the cosine of the angle between the vectors a_i and b_i, the last assertion follows. □

Remark 5.2.10 The matrix D in Theorem 5.2.9 will be called the *reduction matrix* of the cone C. In fact, it is at the same time also the reduction matrix of the polar cone $\tilde{C}$. The diagonal entries d_i of D will be called *reduction parameters*. We shall find their geometric properties later (in Corollary 5.2.13).

The simplicial n-cone has, of course, not only $(n-1)$-dimensional faces, but faces of all dimensions between 1 and $n-1$. Such a face is simply generated by a subset of the generators and also forms a simplicial cone. Its Gramian is a principal submatrix of the Gramian of the original n-cone. We can thus ask about the geometric meaning of the corresponding polar cone of the face and its relationship to the polar cone of the original simplex. By Theorem 5.2.1 we obtain the following.

Theorem 5.2.11 *Suppose that a simplicial n-cone C is generated by the vectors $a_1, \dots, a_n$ and the polar cone $\hat{C}$ by the vectors $b_1, \dots, b_n$ which complete the a_is to biorthogonal bases. Let F be the face of C generated by $a_1, \dots, a_k$, $1 \le k \le n-1$. Then the polar cone $\hat{F}$ of F is generated by the orthogonal projections of the first k vertex halflines of $\hat{C}$ on the linear space spanned by the vectors $a_1, \dots, a_k$.*

There are some distinguished halflines of the n-cone C. In the following theorem, the spherical distance of a halfline h from a hyperplane is the smallest angle the halfline h spans with the halflines in the hyperplane.

Theorem 5.2.12 *There is exactly one halfline h, which has the same spherical distance φ to all generating vectors of the cone C. The positively homogeneous coordinates of h are given by $G(C)^{-1}e$, where $e = [1, 1, \dots, 1]^T$ and $G(C)$ is the matrix (5.5). The angle is*

$$\varphi = \arccos \frac{1}{\sqrt{e^T [G(C)]^{-1} e}}. \tag{5.7}$$

The halfline h also has the property that it has the same spherical distance to all $(n-1)$-dimensional faces of the polar cone. This distance is

$$\psi = \arcsin \frac{1}{\sqrt{e^T [G(C)]^{-1} e}}, \qquad 0 < \psi < \frac{\pi}{2}. \tag{5.8}$$

Proof. Suppose that a unit vector c spans with each of the unit vectors a_i the same acute angle φ. Then $\cos\varphi = \langle a_i, c\rangle$ for all i.

If $c = \sum_i \gamma_i a_i$, and $\gamma = [\gamma_1, \dots, \gamma_n]^T$, we obtain

$$G(C)\gamma = e \cos\varphi, \tag{5.9}$$

so that indeed γ is a positive multiple of $G(C)^{-1}e$. The converse also holds.

Since c is a unit vector, we obtain by (5.9) that

$$\begin{aligned} 1 &= \langle c, c\rangle \\ &= \sum_{i,k} \gamma_i\gamma_k\langle a_i, a_k\rangle \\ &= \gamma^T G(C)\gamma \\ &= e^T[G(C)^{-1}]e\cos^2\varphi \end{aligned}$$

so that (5.7) holds. Since c spans with a_i the acute angle φ, it spans with the $(n-1)$-dimensional face of the polar cone, generated by all the b_j, $i \neq j$, the same angle complementing φ to $\frac{\pi}{2}$. Thus the spherical distance ψ of h to the faces of the polar cone is the same and satisfies (5.8). □

We could thus speak about the *circumscribed circular cone* with axis h of the cone C and about the *inscribed circular cone*, in this case of the polar cone $\tilde{C}$. Of course, we can exchange the roles of C and $\tilde{C}$. If C is a general circular cone, we shall call the angle between the *axis* of the cone and any boundary vector the *opening angle* of the cone. We summarize in the following result:

Corollary 5.2.13 *Let C be a simplicial n-cone with Gramian $G(C)$. Then the opening angle of the circumscribed circular cone of C is*

$$\varphi = \arccos \frac{1}{\sqrt{e^T[G(C)]^{-1}e}},$$

and the axis is generated by the vector $c = \sum_k \gamma_k a_k$ for $\gamma = [\gamma_1, \ldots, \gamma_n]^T$ satisfying

$$\gamma = [G(C)]^{-1}e.$$

The opening angle of the inscribed circular cone of C is

$$\rho = \arcsin \frac{1}{\sqrt{e^T D^{-1}G(C)D^{-1}e}}, \tag{5.10}$$

where D is the reduction matrix of C, and the axis is generated by the vector

$$v = \sum_k d_k^{-1}a_k, \tag{5.11}$$

where the d_ks are the reduction parameters.

Proof. The first part is in Theorem 5.2.12. Since the normalized Gramian of the polar cone $\tilde{C}$ is $D[G(C)]^{-1}D$ and the polar cone of $\tilde{C}$ is C, (5.8) applied to $\tilde{C}$ yields (5.10) and (5.11). □

Remark 5.2.14 However, one must bear in mind that there exist circumscribed and inscribed cones of the conjugate cones, too. Thus there exist in general 2^n circumscribed circular cones in this more general sense (or,

2^{n-1} double-cones), and the same number of inscribed circular cones or double-cones.

Remark 5.2.15 For the purpose of this section, we introduce two notions. If C is an n-cone and H a hyperplane which has a proper point in common with every generating ray of C, distinct from the vertex of C, then these intersection points and the vertex of C form vertices of an n-simplex. We then call it the *cut-off n-simplex* of C for obvious reasons. Conversely if Σ is an n-simplex and V is its vertex, then we call the *vertex-cone* of Σ at V the n-cone generated by the vectors of the edges of Σ starting at V.

We can now prove a theorem which will enable us to apply the results in the previous chapters.

Theorem 5.2.16 *For every simplicial n-cone C, there exists a cut-off n-simplex of C such that all interior dihedral angles of the simplex at the added boundary hyperplane are acute.*

Proof. Suppose that $a_1, \ldots, a_n$ are the unit generators of C, and $G(a)$ is its Gramian. By a well-known theorem (Appendix, Corollary A.3.13), there exists a positive vector $c = [c_i]^T$ such that $G(a)c = u$ is positive as well. Define now a halfspace H^+ by $\min_i c_i - \langle \sum_i c_i a_i, x\rangle \geq 0$. The boundary hyperplane H cuts each of the halflines λa_k, $\lambda > 0$, in the point determined by $\lambda = \min_i c_i/u_i$, where u_i is the ith coordinate of u. The intersection $C \cap H^+$ is thus an n-simplex Σ, and the vector $b_{n+1} = \sum_i c_i a_i$ is the $(n+1)$th outer normal, with the remaining normals being $-b_1, \ldots, -b_n$. Since the inner product between b_{n+1} and each outer normal $-b_k$ is $-c_k$, i.e. negative, all the interior angles in Σ at H are acute. □

In the following, assign to the cone C a cut-off n-simplex Σ by choosing the vertices of Σ as points on the generating halflines in a unit distance from the vertex of C denoted as A_{n+1}. Now we can define the *isogonal correspondence* among nb-halflines of C, i.e. halflines not contained in any $(n-1)$-dimensional face of C, as follows. If h_1 is such a halfline, choose on h_1 a point P not contained in the $(n-1)$-dimensional face ω_{n+1} of Σ opposite to A_{n+1}. By Theorem 2.2.10, there exists a unique point Q which is isogonally conjugate to P with respect to the simplex Σ. We then call the halfline h_2 originating at A_{n+1} and containing Q, the *isogonally conjugate halfline* to h_1. It is immediate from Theorem 2.2.4 that the isogonally conjugate halfline to h_2 with respect to Σ is h_1. Let us show that h_2 is independent of the choice of the simplex Σ. Indeed, as was proved in Theorem 2.2.11, the point Q is the center of a hypersphere circumscribed to the n-simplex with vertices in n points symmetric to P with respect to each of the $(n-1)$-dimensional faces of C, together with the point Z symmetric to P with respect to ω_{n+1}. Since A_{n+1} has the same distance to all the mentioned n points, the line $A_{n+1}Q$ containing h_2 is the locus of all points having the same distance to these n points, and this line does not depend on the position of Z. We thus proved the following:

Theorem 5.2.17 *The isogonally conjugate halfline h_2 to the halfline h_1 is the locus of the other foci of rotational hyperquadrics which are inscribed into the cone C and have one focus at h_1. In addition, whenever we choose two $(n-1)$-dimensional faces F_1 and F_j of C, then the two hyperplanes in the pencil $\alpha_1 F_1 + \alpha_2 F_2$ (in the clear sense) passing through h_1 and h_2 are* symmedians, *i.e. they are symmetric with respect to the axes of symmetry of F_1 and F_2.*

Remark 5.2.18 The case that the two halflines coincide clearly happens if and only if this rotational hyperquadric is a hypersphere. The halflines then coincide with the axis of the inscribed circular cone of C (cf. Theorem 5.2.12).

Observe that the isogonal correspondence between the halflines in C is at the same time a correspondence between two halflines in the polar cone $\bar{C}$, namely such that they are not *nb*-halflines with respect to $\bar{C}$, which means that they are not orthogonal to any $(n-1)$-dimensional face of C. If we now exchange the role of C and $\bar{C}$, we obtain another correspondence in C between halflines of C not orthogonal to any $(n-1)$-dimensional face of C. In this case, two such halflines coincide if and only if they coincide with the halfaxis of the circumscribed circular cone of C.

The notion of hyperacute simplexes plays an analogous role in simplicial cones. A simplicial n-cone C generated by the vectors a_i is called *hyperacute* if none of the angles between the $(n-1)$-dimensional faces of C is obtuse. By Theorem 5.2.16, we then have:

Theorem 5.2.19 *If C is a hyperacute n-cone, then there exists a cut-off n-simplex of C which is also hyperacute.*

By Theorem 3.3.1, we immediately obtain:

Corollary 5.2.20 *Every face (with dimension at least two) of a hyperacute n-cone is also hyperacute.*

The property of an n-cone to be hyperacute is, of course, equivalent to the condition that for the polar cone $\langle b_i, b_j \rangle \leq 0$ for all i, j. To formulate consequences, it is advantageous to define an n-cone as *hypernarrow* (respectively, *hyperwide*) if any two angles between the generators are acute or right (respectively, obtuse or right). We then have:

Theorem 5.2.21 *An n-cone is hyperacute if and only if its polar cone is hyperwide.*

By Theorem 5.2.19, the following holds:

Theorem 5.2.22 *A hyperacute n-cone is always hypernarrow.*

Remark 5.2.23 Of course, the converse of Theorem 5.2.22 does not hold for $n \geq 3$.

Analogously, we can say that a simplicial cone generated by the vectors a_i is *hyperobtuse* if none of the angles between a_i and a_j is acute.

The following are immediate:

Theorem 5.2.24 *If a simplicial cone C is hypernarrow (respectively, hyperwide), then every face of C is hypernarrow (respectively, hyperwide) as well.*

Theorem 5.2.25 *If a simplicial cone is hyperwide, then its polar cone is hypernarrow.*

Also, the following is easily proved.

Theorem 5.2.26 *The angle spanned by any two rays in a hypernarrow cone is always either acute or right.*

Analogous to the point case, we can define orthocentric cones. First, we define the *altitude hyperplane* as the hyperplane orthogonal to an $(n-1)$-dimensional face and passing through the opposite vertex halfline. An n-cone C is then called orthocentric if there exist n altitude hyperplanes which meet in a line; this line will be called the *orthocentric line.*

Remark 5.2.27 The altitude hyperplane need not be uniquely defined if one of the vertex halflines is orthogonal to all the remaining vertex halflines. It can even happen that each of the vertex halflines is orthogonal to all the remaining ones. Such a totally orthogonal cone is, of course, also considered orthocentric. We shall, however, be interested in cones – we shall call them *usual* – in which no vertex halfline is orthogonal to *any* other vertex halfline, thus also not to the opposite face, and for simplicity, that the same holds for the polar cone. Observe that such a cone has the property that the polar cone has no vertex halfline in common with the original cone.

Theorem 5.2.28 *Any usual simplicial* 3*-cone is orthocentric.*

Proof. Suppose that C is a usual 3-cone generated by vectors a_1, a_2, and a_3. Let b_1, b_2, and b_3 complete the generating vectors to biorthogonal bases. We shall show that the vector

$$s = \frac{1}{\langle a_2, a_3\rangle} b_1 + \frac{1}{\langle a_3, a_1\rangle} b_2 + \frac{1}{\langle a_1, a_2\rangle} b_3$$

generates the orthocentric line. Let us prove that the vector s is a linear combination of each pair a_i, b_i, for $i = 1, 2, 3$. We shall do that for $i = 1$. If the symbol $[x, y, z]$, where x, y, and z are vectors in E_3, means the 3×3 matrix of cartesian coordinates of these vectors form the product $[a_1, a_2, a_3]^T[b_1, a_1, s]$. We obtain for the determinants

$$\det[a_1, a_2, a_3]^T \det[b_1, a_1, s] = \det \begin{bmatrix} \langle a_1, b_1\rangle & \langle a_1, a_1\rangle & \langle a_1, s\rangle \\ \langle a_2, b_1\rangle & \langle a_2, a_1\rangle & \langle a_2, s\rangle \\ \langle a_3, b_1\rangle & \langle a_3, a_1\rangle & \langle a_3, s\rangle \end{bmatrix}.$$

The determinant on the right-hand side is

$$\det \begin{bmatrix} 1 & \langle a_1, a_1\rangle & \frac{1}{\langle a_2, a_3\rangle} \\ 0 & \langle a_2, a_1\rangle & \frac{1}{\langle a_1, a_3\rangle} \\ 0 & \langle a_3, a_1\rangle & \frac{1}{\langle a_2, a_1\rangle} \end{bmatrix},$$

which is zero. The same holds for $i = 2$ and $i = 3$. Thus, the plane containing linearly independent vectors a_1 and s contains also the vector b_1 so that it is orthogonal to the plane generated by a_2 and a_3. It is an altitude plane containing a_1. Therefore, s is the orthocentric line. □

Returning to Theorem 4.2.4, we can prove the following:

Theorem 5.2.29 *A usual n-cone, $n \geq 3$, is orthocentric if and only if its Gramian has d-rank one. The polar cone is then also orthocentric.*

Proof. For $n = 3$, the result is correct by Theorem 5.2.28. Suppose the d-rank of the usual n-cone C is one and $n > 3$. The Gramian $G(C)$ thus has the form $G(C) = D + \mu uu^T$, where D is a diagonal matrix, u is a column vector with all entries u_i different from zero, and μ is a real number different from zero. We shall show that the line s generated by the vector

$$v = \sum_i u_i b_i \tag{5.12}$$

satisfying $\langle v, a_i\rangle = u_i$ is then contained in all the two-dimensional planes P_i, each generated by the vectors a_i and b_i, $i = 1, \ldots, n$, where the b_is complete the a_is to biorthogonal bases. Indeed, let $k \in \{1, \ldots, n\}$, and multiply the nonsingular matrix $A = [a_1, \ldots, a_n]^T$ by the $n \times 3$ matrix $B_k = [b_k, a_k, v]$. We obtain

$$AB_k = \begin{bmatrix} \langle a_1, b_k\rangle & \langle a_1, a_k\rangle & \langle a_1, v\rangle] \\ \langle a_2, b_k\rangle & \langle a_2, a_k\rangle & \langle a_2, v\rangle] \\ \vdots & \vdots & \vdots \\ \langle a_n, b_k\rangle & \langle a_n, a_k\rangle & \langle a_n, v\rangle] \end{bmatrix}.$$

This matrix has rank two since in the first column there is just one entry, the kth, different from zero, and the second column is – except for the kth entry – a multiple of u, and the same holds for the third column. Thus s is contained in each altitude hyperplane containing a vertex line and orthogonal to the opposite $(n-1)$-dimensional face; it is an orthocentric line of C, and the only one since a_k and b_k are for each k linearly independent.

Conversely, take in an n-cone C, a line s to be an orthocentric line generated by a nonzero vector w. Then w is contained in every plane P_k generated by the pair a_k and b_k as above. Thus the corresponding $n \times 3$-matrix $B_k = [b_k, a_k, w]$ has rank two for each k, which implies, after multiplication by A as above, that each off-diagonal entry $\langle a_i, a_k \rangle$ is proportional to $\langle a_i, w \rangle$, $i \neq k$. Since we assume that C is usual and $n \geq 3$, it cannot happen that the last column has all entries except the kth equal to zero. In such a case, w would be proportional to b_k, and b_k would have to be linearly dependent on a_j and b_j for some $j \neq k$. The inner product of a_i, where i is different from both j and k, with a_j would then be equal to zero. Therefore, there exist constants c_k different from zero such that $\langle a_i, a_k \rangle = c_k \langle a_i, w \rangle$ for all i, k, $i \neq k$. Thus the d-rank of $G(C)$ is one.

The fact that the polar cone is also orthocentric follows from the fact that the orthocentric line is symmetrically defined with respect to both parts of the biorthogonal bases a_i and b_i. □

Theorem 5.2.30 *If a usual cone C is orthocentric, then every face of C is also orthocentric.*

Proof. It follows from the fact that the properties of being usual and being Gramian of d-rank one are hereditary. □

The change of signs of the generators a_i does not change the property that the d-rank is one, and the usual property of a cone remains. Therefore, we have also

Theorem 5.2.31 *If a usual cone is orthocentric, then all its conjugate cones are orthocentric as well. The orthocentric lines of the conjugate cones are related to the orthocentric line of the original n-cone by conjugacy, in this case by multiplication of some of the coordinates with respect to the basis b_i by minus one.*

Proof. The first part follows from the fact that the change of signs of the generators a_i does not change the property that the d-rank is one, and also the property of the cone to be usual remains. The second part is a consequence of (5.12). □

Remark 5.2.32 The existence of an orthocentric n-cone for any $n \geq 3$ follows from the fact that for every positive definite $n \times n$ matrix there exist n linearly independent vectors in E_n whose Gram matrix is the given matrix. If we choose a diagonal matrix D with positive diagonal entries and a column vector u with positive entries, the matrix $D + uu^T$ will certainly correspond to a usual orthocentric cone.

Analogous to the properties of the orthocentric n-simplex, the following holds:

Theorem 5.2.33 *Suppose that $a_1, \ldots, a_n$ generate a usual orthocentric n-cone with orthocentric line generated by the vector a_0. Then each n-cone generated by any n-tuple from the vectors $a_0, a_1, \ldots, a_n$ is usual and orthocentric and the remaining vector generates the corresponding orthocentric line.*

Proof. Let C be the usual cone generated by $a_1, \ldots, a_n$, and let a_0 be a generator of the orthocentric line. We shall show that whenever we choose a_r and a_s, $r \neq s$, $r, s \in \{0, 1, \ldots, n\}$, then there is a nonzero vector v_{rs} in the plane P_{rs} generated by a_r and a_s, which is orthogonal to the hyperplane H_{rs} generated by all a_ts, $r \neq t \neq s$. If one of the indices r, s is zero, the assertion is true: if the other index is, say, one, the vector b_1 is linearly dependent on a_0 and a_1. Thus let both indices r, s be different from zero; let, say, $r = 1$, $s = 2$. The nonzero vector $v_{12} = \langle b_2, a_0 \rangle b_1 - \langle b_1, a_0 \rangle b_2$ is orthogonal to a_0, a_3, ..., a_n, thus to H_{12}. Let us show that $v_{12} \in P_{12}$. We know that a_0 is a linear combination both of a_1, b_1 and of a_2, b_2: $a_0 = \alpha_1 a_1 + \beta_1 b_1$, $a_0 = \alpha_2 a_2 + \beta_2 b_2$. Thus $\langle b_2, a_0 \rangle = \beta_1 \langle b_1, b_2 \rangle$, $\langle b_1, a_0 \rangle = \beta_2 \langle b_1, b_2 \rangle$, $\alpha_1 a_1 - \alpha_2 a_2 = \beta_2 b_2 - \beta_1 b_1$ so that

$$\frac{1}{\langle b_1, b_2 \rangle} v_{12} = \alpha_2 a_2 - \alpha_1 a_1.$$

□

Observe that the notion of the vertex-cone defined in Remark 5.2.15 enables us to formulate a connection between orthocentric simplexes and orthocentric cones.

Theorem 5.2.34 *If an orthocentric n-simplex is different from the right, then its every vertex-cone is orthocentric and usual. In addition, the line passing through the vertex and the orthocenter is then the orthocentric line of the vertex cone.*

We could expect that the converse is also true. In fact, it is true for so-called *acute orthocentric cones*, i.e. orthocentric cones all of whose interior angles are acute. In this case, there exists an *orthocentric ray* (as a part of the orthocentric line) which is in the interior of the cone.

Then, the following holds:

Theorem 5.2.35 *If C is a usual acute orthocentric n-cone, then there exists a cut-off n-simplex of C which is acute orthocentric and different from the right.*

Proof. Choose a proper point P on the orthocentric ray of C, different from the vertex V of C. If a_i is a generating vector of C, and b_i is the corresponding vector of the biorthogonal basis, then by the proof of Theorem 5.2.29, P, a_i, and b_i are in a plane. Since a_i and b_i are linearly independent, there exists on the *line* $V + \lambda a_i$, a point X_i of the form $X_i = P + \xi b_i$. Let us show that λ is positive. We have $P - V = \lambda a_i - \xi b_i$. Let $j \neq i$ be another index. Then $\langle P - V, a_j \rangle \geq 0$ by Theorem 5.2.26 so that $\lambda \langle a_i, a_j \rangle \geq 0$. Since λ cannot be zero, it is positive. Thus X_i is on the ray $V + \lambda a_i$ for λ positive, and this holds for all i. The points X_i, together with the vertex V, form vertices of the acute orthocentric n-simplex. □

We can ask now what happens if we have more generators of a cone in E_n than n. As before, the cone will be defined as the set of all nonnegative linear combinations of the given vectors. We shall suppose that this set does not contain a line and has dimension n in the sense that it is not contained in any E_k with $k < n$. The resulting cone will then be called *simple*. The following is almost immediate:

Theorem 5.2.36 *Let $\mathcal{S} = \{a_1, \ldots, a_m\}$ be a system of vectors in E_n. Then $\mathcal{S}$ generates a simple cone C if and only if*

(i) *C is n-dimensional;*
(ii) *the zero vector can be expressed as a nonnegative combination of the vectors a_i, thus in the form $\sum_i \alpha_i a_i$, where all the αs are nonnegative, only if all the αs are zero.*

Another situation can occur if at least one of the vectors a_i is itself a nonnegative combination of other vectors of the system. We then say that such vector is *redundant*. A system of vectors in E_n will be called, for the moment, *pure* if it generates a simple cone and none of the vectors of the system is redundant. Returning now to biorthogonal systems, we have:

Theorem 5.2.37 *Let $\mathcal{S}$ be a pure system of vectors in E_n. Then the system $\hat{\mathcal{S}}$ which is the biorthogonal system to $\mathcal{S}$ is also pure.*

Proof. This follows from the fact that the vectors of the system $\hat{\mathcal{S}}$ satisfy the same linear relations as the corresponding vectors in $\mathcal{S}$. □

5.3 Regular simplicial cones

For completeness, we add a short section on simplicial cones which possess the property that the angles between any two generating halflines are the same. We shall call such cones *regular*. The sum of the generating unit vectors will be called the *axis* of the cone, the common angle of the generating vectors will be the *basic angle*, and the angle between the axis and the generating vectors will be the *central angle*.

Theorem 5.3.1 *Suppose that Σ is a regular simplicial n-cone. Denote by α its basic angle, and by ω its central angle. Then the polar cone Σ' of Σ is also regular and the following relations hold between its basic angle α' and central angle ω'*

$$\cos\alpha \leq \frac{1}{n},$$

$$\cos^2\omega = \frac{1}{n}(1-(n-1)\cos\alpha),$$

$$\cos\alpha' = -\frac{\cos\alpha}{1-(n-2)\cos\alpha},$$

$$\cos^2\omega' = \frac{1-\cos^2\omega}{1+n(n-2)\cos^2\omega}.$$

The common interior angle φ between any two faces of Σ satisfies, of course, $\varphi = \pi - \alpha'$, and similarly for the polar cone, $\varphi' = \pi - \alpha$.

Proof. The Gram matrices of the cones Σ and Σ' have the form $I - kE$ and $I - k'E$, where E is the matrix of all ones and, since these matrices have to be positive definite and mutual inverses, $k < \frac{1}{n}$, $k' = -\frac{k}{1-kn}$. Thus $\cos\alpha = \frac{k}{1-k}$; if the a_is are the unit generating vectors of Σ, $c = \sum_i a_i$ is the generating vector of the axis. Then $\langle a_i, a_i\rangle = 1-k$, $\langle a_i, a_j\rangle = -k$, $\cos^2\omega = \frac{(\langle c,a_i\rangle)^2}{\langle c,c\rangle\langle a_i,a_i\rangle}$. Simple manipulations then yield the formulae above. □

Remark 5.3.2 It is easily seen that a regular cone is always orthocentric; its orthocentric line is, of course, the axis.

Remark 5.3.3 Returning to the notion of simplexes with a principal point, observe that every n-simplex with principal point for which the coefficient α is positive can be obtained by choosing a regular $(n+1)$-cone C and a hyperplane H not containing its vertex. The intersection points of the generating halflines of C with H will be the vertices of such an n-simplex.

5.4 Spherical simplexes

If we restrict every nonzero vector in E_n to a unit vector by appropriate multiplication by a positive number, we obtain a point on the unit sphere S_n in E_n. In particular, to a simplicial n-cone there corresponds a *spherical n-simplex* on S_n. For $n = 3$, we obtain a *spherical triangle.* In the second definition, we have to use hemispheres instead of halfspaces, so that the given n-simplex can be defined as the intersection of the n hemispheres, each containing $n-1$

of the given points on the boundary and the remaining point as an interior point.

Such a general hemisphere corresponds to a unit vector orthogonal to the boundary hyperplane and contained in the hemisphere, so-called *polar vector*, and conversely, to every unit vector, or, a point on the hypersphere S_n, corresponds to a unique *polar* hemisphere described. It is immediate that the polar hemisphere corresponding to the unit vector u coincides with the set of all unit vectors x satisfying $\langle u, x\rangle \geq 0$. We can then define the *polar spherical n-simplex* $\tilde{\Sigma}$ to the given spherical n-simplex Σ generated by the vectors a_i as the intersection of all the polar hemispheres corresponding to the vectors a_i.

It is well known that the *spherical distance* of two points a, b can be defined as $\arccos |\langle a, b\rangle|$ and this distance satisfies the triangular inequality among points in a hemisphere. It is also called the *spherical length* of the *spherical arc ab*. By Theorem 5.2.7, the spherical simplex is by the lengths of all the arcs a_ia_j determined up to the position on the sphere. The lengths of the arcs between the vertices of the polar simplex correspond to the *interior angles* of the original simplex in the sense that they complete them to π.

The matricial approach to spherical simplexes allows us – similarly as for simplicial cones – to consider also qualitative properties of the angles. We shall say that an arc between two points a and b of S_n is *small* if $\langle a, b\rangle > 0$, *medium* if $\langle a, b\rangle = 0$, and *large* if $\langle a, b\rangle < 0$. We shall say that an n-simplex is *small* if each of the arcs is small or medium, and *large* if each of its arcs is large or medium. Finally, we say that an n-simplex is *hyperacute* if each of the interior angles is acute or right.

The following is trivial:

Theorem 5.4.1 *If a spherical n-simplex is small (respectively, large), then all its faces are small (respectively, large) as well.*

Theorem 5.4.2 *The polar n-simplex of a spherical n-simplex Σ is large if and only if Σ is hyperacute.*

Less immediate is:

Theorem 5.4.3 *If a spherical n-simplex is hyperacute, then it is small.*

Proof. By Theorems 5.2.7, 5.2.8, and 5.4.2, the Gramian of the polar of a hyperacute spherical n-simplex is an M-matrix. Since the inverse of an M-matrix is a nonnegative matrix by Theorem A.3.2, the result follows. □

Theorem 5.4.4 *If a spherical n-simplex is hyperacute, then all its faces are hyperacute.*

Proof. If C is hyperacute, then again the Gramian of the polar cone is an M-matrix. The inverse of the principal submatrix corresponding to the face is thus a Schur complement of this Gramian. By Theorem A.3.3 in the Appendix, it is again an M-matrix so that the polar cone of the face is large. By Theorem 5.4.2, the face is hyperacute. □

We can repeat the results of Section 2 for spherical simplexes. In particular, the results on circumscribed and inscribed circular cones (in the spherical case, hyperspheres with radius smaller than one on the unit hypersphere) will be valid. Also, conjugacy and the notion that an n-simplex is usual can be defined. There is also an analogy to the isogonal correspondence for the spherical simplex as mentioned in Theorem 5.2.17, as well as isogonal correspondence with respect to the polar simplex.

Also, the whole theory about the orthocentric n-cones can be used for spherical simplexes. Let us return to Theorem 5.2.9, where the Gramian of normalized vectors of two biorthogonal bases was described. In [14], it was proved that a necessary and sufficient condition for the diagonal entries a_{ii} of a positive definite matrix A and the diagonal entries α_{ii} of the inverse matrix A^{-1} is

$$a_{ii}\alpha_{ii} \geq 1 \text{ for all } i;$$

and

$$2\max_i(\sqrt{a_{ii}\alpha_{ii}} - 1) \leq \sum_i(\sqrt{a_{ii}\alpha_{ii}} - 1). \tag{5.13}$$

If we apply this result to the Gramians of C and $\hat{C}$ (multiplied by D^{-1} from both sides), we obtain that the diagonal entries d_i^{-1} of D^{-1} satisfy, in addition to $0 < d_i \leq 1$, the inequality

$$2\max_i(d_i^{-1} - 1) \leq \sum_i(d_i^{-1} - 1). \tag{5.14}$$

Observing that d_i is by (5.6) the cosine of the angle φ_i between a_i and b_i, it follows that the modulus of $\frac{\pi}{2} - \varphi_i$ is the angle ψ_i of the altitude between the vector a_i and the opposite face of C. Since (5.14) means

$$2\max_i(\sec\varphi_i - 1) \leq \sum_i(\sec\varphi_i - 1),$$

we obtain the following.

Theorem 5.4.5 *A necessary and sufficient condition that the angles ψ_i, $i = 1, \ldots, n$, be angles of the altitudes in a spherical n-simplex, is*

$$2\max_i(\csc\psi_i - 1) \leq \sum_i(\csc\psi_i - 1). \tag{5.15}$$

Remark 5.4.6 In [14], a necessary and sufficient condition for a positive definite matrix A was found in order that equality in (5.13) is attained. It can be shown that the corresponding geometric interpretation for a spherical

n-simplex Σ satisfying equality in (5.15) is that the polar n-simplex $\tilde{\Sigma}$ is symmetric to Σ with respect to a hyperplane. In addition, both simplexes are orthocentric.

We make a final comment on the spherical geometry. As we saw, spherical geometry is in a sense richer than the Euclidean since we can study the polar objects. On the other hand, we are losing one dimension (in E_3, we can visualize only spherical triangles and not spherical tetrahedrons). In the Euclidean geometry, we have the centroid which we do not have in spherical geometry, etc.

5.5 Finite sets of points

As we saw in Chapter 1, we can study problems in $\mathcal{E}_n$ using the barycentric coordinates with respect to a simplex, i.e. a distinguished set of $n+1$ linearly independent points. We can ask the question of whether we can do something analogous in the case that we have more than $n+1$ distinguished points in $\mathcal{E}_n$.

In fact, we can again define barycentric coordinates.

Suppose that $A_1, A_2, \ldots, A_m$ are points in $\mathcal{E}_n$, $m > n+1$. A linear combination $\sum_1^m \alpha_i A_i$ has again geometric meaning in two cases. If the sum $\sum \alpha_i = 1$, we obtain a point; if $\sum \alpha_i = 0$, the result is a vector. Of course, all such combinations describe just the smallest linear space of $\mathcal{E}_n$ containing the given points.

Analogously to the construction in Chapter 1, we can define the *homogeneous barycentric coordinates* as follows.

A linear combination $\sum_i \beta_i A_i$ is a vector if $\sum_i \beta_i = 0$; if $\sum_i \beta_i \neq 0$, then it is the point $\sum_i \gamma\beta_i A_i$, where $\gamma = (\sum_i \beta_i)^{-1}$. We have thus a correspondence between the points and vectors in $\bar{\mathcal{E}}_n$ and the $(m-1)$-dimensional projective space P_{m-1}. In this space, we can identify a linear subspace formed by the linear dependencies among the points A_i. We shall illustrate the situation by an example.

Example 5.5.1 Let $A = (0,0)$, $B = (1,0)$, $C = (1,1)$, and $D = (0,1)$ be four such points in $\mathcal{E}_2$ in the usual coordinates. The point $\left(\frac{1}{2}, \frac{1}{2}\right)$ has the expression $\frac{1}{4}A + \frac{1}{4}B + \frac{1}{4}C + \frac{1}{4}D$, but also $\frac{1}{2}A + \frac{1}{2}C$. This is, of course, caused by the fact that there is a linear dependence relation $A-B+C-D = 0$ among the given points. The projective space mentioned above is three-dimensional, but there is a plane P_2 with the equation $x_1 - x_2 + x_3 - x_4 = 0$ having the property that there is a one-to-one correspondence between the points of the plane $\bar{\mathcal{E}}_2$, i.e. the Euclidian plane $\mathcal{E}_2$ completed by the points at infinity, and the plane P_2. In particular, the improper points of $\bar{\mathcal{E}}_2$ correspond to the points in P_2 contained in the plane $x_1 + x_2 + x_3 + x_4 = 0$.

The squares of the mutual distances of the given points form the matrix

$$M = \begin{bmatrix} 0 & 1 & 2 & 1 \\ 1 & 0 & 1 & 2 \\ 2 & 1 & 0 & 1 \\ 1 & 2 & 1 & 0 \end{bmatrix}.$$

Observe that the bordered matrix (as was done in Chapter 1, Corollary 1.4.3)

$$M_0 = \begin{bmatrix} 0 & 1 & 1 & 1 & 1 \\ 1 & 0 & 1 & 2 & 1 \\ 1 & 1 & 0 & 1 & 2 \\ 1 & 2 & 1 & 0 & 1 \\ 1 & 1 & 2 & 1 & 0 \end{bmatrix}$$

is singular since $M_0\,[0, 1, -1, 1, -1]^T = 0$.

On the other hand, if a vector $x = [x_1, x_2, x_3, x_4]^T$ satisfies $\sum x_i = 0$, then $x^T\,M\,x$ is after some manipulations (subtracting $(x_1 + x_2 + x_3 + x_4)^2$, etc.) equal to $-(x_1 + x_3)^2 - (x_2 + x_4)^2$, thus nonpositive and equal to zero if and only if the vector is a multiple of $[1, -1, 1, -1]^T$.

Let us return to the general case. The squares of the mutual distances of the points A_i form the matrix $M = [|A_i - A_j|^2]$. As in the case of simplexes, we call it the *Menger matrix* of the (ordered) system of points. The following theorem describes its properties.

Theorem 5.5.2 *Let $\mathcal{S} = \{A_1, A_2, \ldots, A_m\}$ be an ordered system of points in $\mathcal{E}_n$, $m > n+1$. Denote by $M = [m_{ij}]$ the Menger matrix of $\mathcal{S}$, $m_{ij} = |A_i - A_j|^2$, and by M_0 the bordered Menger matrix*

$$M_0 = \begin{bmatrix} 0 & e^T \\ e & M \end{bmatrix}. \tag{5.16}$$

Then:

(i) $m_{ii} = 0$, $m_{ij} = m_{ji}$;

(ii) *whenever $x_1, \ldots, x_m$ are real numbers satisfying $\sum_i x_i = 0$, then*

$$\sum_{i,j=1}^{m} m_{ij} x_i x_j \le 0;$$

(iii) *the matrix M_0 has rank $s+1$, where s is the maximum number of linearly independent points in $\mathcal{S}$.*

Conversely, if $M = [m_{ij}]$ is a real $m \times m$ matrix satisfying (i), (ii) *and the rank of the corresponding matrix M_0 is $s + 1$, then there exists a system of* *points in a Euclidean space with rank s such that M is the Menger matrix of this ordered system.*

Proof. In the first part, (i) is evident. To prove (ii), choose some orthonormal coordinate system in $\mathcal{E}_n$. Then let $(\overset{k}{a}_1, \dots, \overset{k}{a}_n)$ represent the coordinates of A_k, $k = 1, \dots, m$. Suppose now that $x_1, \dots, x_m$ is a nonzero m-tuple satisfying $\sum_{i=1}^{m} x_i = 0$. Then

$$\begin{aligned}
\sum_{i,k=1}^{m} m_{ik} x_i x_k &= \sum_{i,k=1}^{m} \left(\sum_{\alpha=1}^{n} (\overset{i}{a}_\alpha - \overset{k}{a}_\alpha)^2 \right) x_i x_k \\
&= \sum_{i=1}^{m} \left(\sum_{\alpha=1}^{n} \overset{i}{a}{}_\alpha^2 \right) x_i \sum_{k=1}^{m} x_k + \sum_{i=1}^{m} x_i \sum_{k=1}^{m} \left(\sum_{\alpha=1}^{n} \overset{k}{a}{}_\alpha^2 \right) x_k - \\
&\quad -2 \sum_{i,k=1}^{m} \sum_{\alpha=1}^{n} \overset{i}{a}_\alpha \overset{k}{a}_\alpha x_i x_k \\
&= -2 \sum_{\alpha=1}^{n} \left(\sum_{k=1}^{m} \overset{k}{a}_\alpha x_k \right)^2 \leq 0.
\end{aligned}$$

Suppose now that the rank of the system is s. Then the matrix

$$\begin{bmatrix} \overset{1}{a}_1 & \dots & \overset{1}{a}_n & 1 \\ \overset{2}{a}_1 & \dots & \overset{2}{a}_n & 1 \\ \vdots & \ddots & \vdots & \vdots \\ \overset{m}{a}_1 & \dots & \overset{m}{a}_n & 1 \end{bmatrix}$$

has rank s. Without loss of generality, we can assume that the first s rows are linearly independent so that each of the remaining $m - s$ rows is linearly dependent on the first s rows. The situation is reflected in the extended Menger matrix M_0 from (5.16) as follows: the matrix $\hat{M}_0$ in the first $s + 1$ rows and columns is nonsingular by Corollary 1.4.3 since the first s points form an $(s-1)$-simplex and this matrix is the corresponding extended Menger matrix. The rank of the matrix M_0 is thus at least $s + 1$. We shall show that each of the remaining columns of M_0, say, the next of which corresponds to the point A_{s+1}, is linearly dependent on the first $s + 1$ columns. Let the linear dependence relation among the first $s + 1$ points A_i be

$$\gamma_1 A_1 + \dots + \gamma_s A_s + \gamma_{s+1} A_{s+1} = 0, \tag{5.17}$$

$$\sum_{i=1}^{s+1} \gamma_i = 0. \tag{5.18}$$

We shall show that there exists a number γ_0 such that the linear combination of the first $s + 2$ columns with coefficients $\gamma_0, \gamma_1, \dots, \gamma_{s+1}$ is zero. Because of

(5.18), it is true for the first entry. Since $|A_p - A_q|^2 = \langle A_p - A_q, A_p - A_q\rangle$, etc., we obtain in the $(i+1)$th entry $\gamma_0 + \sum_{j=1}^{s+1}\gamma_j\langle A_j - A_i, A_j - A_i\rangle$, which, if we consider the points A_p formally as radius vectors from some fixed origin, can be written as $\gamma_0 + \sum_{j=1}^{s+1}\gamma_j\langle A_j, A_j\rangle + \sum_{j=1}^{s+1}\gamma_j\langle A_i, A_i\rangle - 2\sum_{j=1}^{s+1}\gamma_j\langle A_j, A_i\rangle$. The last two sums are equal to zero because of (5.18) and (5.17). The remaining sum does not depend on i so that it can be made zero by choosing $\gamma_0 = -\sum_{j=1}^{s+1}\gamma_j\langle A_j, A_j\rangle$.

It remains to prove the last assertion. It is easy to show that similar to Theorem 1.2.7 we can reformulate the condition (ii) in the following form.

The $(m-1)\times(m-1)$ matrix $C = [c_{ij}]$, $i, j = 1, \ldots, m-1$, where $c_{ij} = m_{im} + m_{jm} - m_{ij}$, is positive semidefinite.

The condition that the rank of M_0 is $s+1$ implies similarly as in Theorem 1.2.4 that the rank of C is s. Thus there exists in a Euclidean space E_s of dimension s a set of vectors $c_1, \ldots, c_{m-1}$, such that $\langle c_i, c_j\rangle = c_{ij}$, $i, j = 1, \ldots, m-1$. Choosing arbitrarily an origin A_m and defining points A_i as $A_m + c_i$, $i = 1, \ldots, m-1$, we obtain a set of points in an s-dimensional Euclidean point space, the Menger matrix of which is M. □

This theorem has important consequences.

Theorem 5.5.3 *The formulae* (1.8) *and* (1.9) *for the inner product of two vectors and for the square of the distance between two points in barycentric coordinates hold in generalized barycentric coordinates as well:*

$$\langle Y - X, Z - X\rangle = -\frac{1}{2}\sum_{i,k=1} m_{ik}\left(\frac{x_i}{\Sigma x_j} - \frac{y_i}{\Sigma y_j}\right)\left(\frac{x_k}{\Sigma x_j} - \frac{z_k}{\Sigma z_j}\right),$$

$$\rho^2(X, Y) = -\frac{1}{2}\sum_{i,k=1} m_{ik}\left(\frac{x_i}{\Sigma x_j} - \frac{y_i}{\Sigma y_j}\right)\left(\frac{x_k}{\Sigma x_j} - \frac{y_k}{\Sigma y_j}\right). \tag{5.19}$$

The summation is over the whole set of points and does not depend on the choice of barycentric coordinates if there is such choice.

Theorem 5.5.4 *Suppose $A_1, \ldots, A_m$ is a system $\mathcal{S}$ of points in $\mathcal{E}_n$ and $M = [m_{ik}]$, the corresponding Menger matrix. Then all the points of $\mathcal{S}$ are on a hypersphere in $\mathcal{E}_n$ if and only if there exists a positive constant c such that for all $x_1, \ldots, x_m$*

$$\sum_{i,k=1}^{m} m_{ik}x_i x_k \leq c\left(\sum_{i=1}^{m} x_i\right)^2. \tag{5.20}$$

If there is such a constant, then there exists the smallest, say c_0, of such constants, and then $c_0 = 2r^2$, where r is the radius of the hypersphere.

Proof. Suppose K is a hypersphere with radius r and center C. Denote by $s_1, \ldots, s_m$ the nonhomogeneous barycentric coordinates of C. Observe that

K contains all the points of $\mathcal{S}$ if and only if $|A_i - C|^2 = r^2$, i.e. by (5.19), if and only if for some $r^2 > 0$

$$\sum_k m_{ik} s_k - \frac{1}{2} \sum_{k,l=1}^{m} m_{kl} s_k s_l = r^2, \qquad i = 1, \dots, m. \tag{5.21}$$

Suppose first that (5.21) is satisfied. Then for every m-tuple $x_1, \dots, x_m$, we obtain

$$\sum_{i,k} m_{ik} x_i s_k = \left(\frac{1}{2} \sum_{k,l=1}^{m} m_{kl} s_k s_l + r^2 \right) \sum_i x_i. \tag{5.22}$$

In particular

$$\sum_{i,k} m_{ik} s_i s_k = 2r^2.$$

Since a proper point $X = (x_1, \dots, x_m)$ belongs to K if and only if $|X - C|^2 = r^2$, i.e.

$$-\frac{\sum_{i,k} m_{ik} x_i x_k}{2(\sum x_i)^2} + \frac{\sum_{i,k} m_{ik} x_i s_k}{\sum x_i} - \frac{\sum_{i,k} m_{ik} s_i s_k}{2} = r^2,$$

we obtain by (5.22) that

$$-\frac{\sum_{i,k} m_{ik} x_i x_k}{2(\sum x_i)^2} + r^2 \geq 0.$$

Thus (even also for $\sum x_k = 0$ by (ii) of Theorem 5.5.2)

$$\sum_{i,k} m_{ik} x_i x_k \leq 2r^2 \left(\sum_k x_k \right)^2.$$

This means that (5.20) holds, and the constant $c_0 = 2r^2$ cannot be improved.

Conversely, let (5.20) hold. Then there exists $c_0 = \max \sum_{i,k=1}^{m} m_{ik} x_i x_k$ over x, for which $\sum_{i=1}^{m} x_i = 1$. Suppose that this maximum is attained at the m-tuple $s = (s_1, \dots, s_m)$, $\sum_k s_k = 1$. Since the quadratic form

$$\left(\sum_{i,k} m_{ik} s_i s_k \right) \left(\sum x_i \right)^2 - \left(\sum s_i \right)^2 \left(\sum_{i,k} m_{ik} x_i x_k \right)$$

is positive semidefinite and attains the value zero for $x = s$, all the partial derivatives with respect to x_i at this point are equal to zero

$$\left(\sum_{i,k} m_{ik} s_i s_k \right) \sum s_j - \left(\sum s_j \right)^2 \sum m_{ij} s_j = 0, \quad i = 1, \dots, m,$$

or, using $\sum s_j = 1$, we obtain the identity

$$\sum_{i,k} m_{ik} x_i s_k = \left(\sum_j x_j\right) \sum_{i,k} m_{ik} s_i s_k.$$

This means that (5.21) holds for $r^2 = \frac{1}{2} \sum_{i,k} m_{ik} s_i s_k$, which is a positive number. □

Remark 5.5.5 Observe that in the case of the four points in Example 5.5.1 the condition (5.20) is satisfied with the constant $c_0 = 1$. Thus the points are on a circle with radius $1/\sqrt{2}$.

Theorem 5.5.6 *Denote by $\mathcal{M}_m$ ($m \geq 2$) the vector space of the $m \times m$ matrices $[a_{ik}]$ such that*

$$a_{ii} = 0, \; a_{ik} = a_{ki}, \qquad i, k = 1, \ldots, m.$$

Then the set of all Menger matrices $[m_{ik}]$, which satisfy the conditions (i) *and* (ii) *from Theorem 5.5.2, forms a convex cone with the zero matrix as the vertex. This cone $\mathcal{S}_m$ is the convex hull of the matrices $A = [a_{ik}]$ of the form*

$$a_{ik} = (c_i - c_k)^2, \quad \sum_{i=1}^{m} c_i = 0, \tag{5.23}$$

with real parameters c_i.

Proof. Let $[m_{ik}] \in \mathcal{S}_m$. By Theorem 5.5.2, there exists in $\mathcal{E}_{m-1}$ a system of points $A_1, \ldots, A_m$ such that

$$m_{ik} = |A_i - A_k|^2, \qquad i, k = 1, \ldots, m.$$

Choose in $\mathcal{E}_{m-1}$ an arbitrary system of cartesian coordinates such that the sums of the first, second, etc., coordinates of the points A_i are equal to zero (we want the centroid of the system $A_1, \ldots, A_m$ to be at the origin).

Then for $A_i = (\overset{i}{a}_1, \ldots, \overset{i}{a}_{m-1})$, not only $\sum_i \overset{i}{a}_\alpha = 0$ for $\alpha = 1, \ldots, m-1$, but also $m_{ik} = \sum_{\alpha=1}^{m-1} (\overset{i}{a}_\alpha - \overset{k}{a}_\alpha)^2$. This means that the point $[m_{ik}]$ is the arithmetic mean of the points of the form (5.23) for

$$\overset{\alpha}{c}_i = \overset{i}{a}_\alpha \sqrt{m-1}, \qquad i = 1, \ldots, m, \; \alpha = 1, \ldots, m-1.$$ □

Remark 5.5.7 In this sense, the ordered systems of m points in $\mathcal{E}_{m-1}$ form a convex cone. This can be used for the study of such systems. We should, however, have in mind that the dimension of the sum of two systems which correspond to systems of smaller rank can have greater rank (however, not more than the sum of the ranks). We can describe geometrically that the sum of two systems (in the above sense) is again a system. In the Euclidean space

$\mathcal{E}_{2m}$, choose a cartesian system of coordinates. In the m-dimensional subspace $\mathcal{E}_{m1}$, which has the last m coordinates zero, construct the system with the first Menger matrix; in the m-dimensional subspace $\mathcal{E}_{m2}$, which has the first m coordinates zero, construct the system with the second Menger matrix. If now A_k is the kth point of the first system and B_k is the kth point of the second system, let C_k be the point whose first m coordinates are those of the point A_k and the last m coordinates those of the point B_k. Then, since

$$|C_i - C_j|^2 = |A_i - A_j|^2 + |B_i - B_j|^2,$$

we have found a system corresponding to the sum of the two Menger matrices (in $\mathcal{E}_{2m}$, but the dimension could be reduced).

Theorem 5.5.8 *The set $\mathcal{P}_{hm}$ of those matrices from $\mathcal{M}_m$ which fulfill the condition* (5.20), *i.e. which correspond to systems of points on a hypersphere, is also a convex cone in $\mathcal{S}_m$.*

In addition, if $\mathcal{A}_1$ is a system with radius r_1 and $\mathcal{A}_2$ a system with radius r_2, then $\mathcal{A}_1 + \mathcal{A}_2$ has radius r fulfilling $r^2 \le r_1^2 + r_2^2$.

Proof. Suppose that both conditions $\sum \overset{1}{m}_{ik} x_i x_k \le 2r_1^2 \ (\sum_i x_i)^2$ and $\sum \overset{2}{m}_{ik} x_i x_k \le 2r_1^2 (\sum_i x_i)^2$ are satisfied; then for $m_{ik} = \overset{1}{m}_{ik} + \overset{2}{m}_{ik}$, $\sum m_{ik} x_i x_k \le (2r_1^2 + 2r_2^2)(\sum_i x_i)^2$. This, together with the obvious multiplicative property by a positive constant, yields convexity. Also, the formula $r^2 \le r_1^2 + r_2^2$ follows by Theorem 5.5.4. □

Theorem 5.5.9 *The set $\mathcal{P}_m$ of those matrices $[a_{ik}]$ from $\mathcal{M}_m$, which satisfy the system of inequalities*

$$a_{ik} + a_{il} \ge a_{kl}, \qquad i, k, l = 1, \dots, m, \tag{5.24}$$

is a convex polyhedral cone.

Proof. This follows from the linearity of the conditions (5.24). □

Remark 5.5.10 Interpreting this theorem in terms of systems of points, we obtain:

The system of all ordered m-tuples of points in $\mathcal{E}_{m-1}$ such that any three of them form a triangle with no obtuse angle, i.e. the set $\mathcal{P}_m \cap \mathcal{S}_m$, forms a convex cone.

Theorem 5.5.11 *Suppose that $c_1, \dots, c_m$ are real numbers, where $\sum_{i=1}^{m} c_i = 0$. The matrix $A = [a_{ik}]$ with entries*

$$a_{ik} = |c_i - c_k|, \qquad i, k = 1, \dots, m, \tag{5.25}$$

is contained in $\mathcal{S}_m$. The set $\hat{\mathcal{P}}_m$, formed as the convex hull of matrices satisfying (5.25), *is a convex cone contained in the intersection $\mathcal{P}_m \cap \mathcal{S}_m$.*

Proof. Suppose $c_{i_1} \leq c_{i_2} \leq \cdots \leq c_{i_m}$. Then the points $A_1, \ldots, A_m$ in $\mathcal{E}_{m-1}$, whose coordinates in some Cartesian coordinate system are

$$\begin{array}{rcllll} A_{i_1} &=& (0, & 0, & \ldots, & 0),\\ A_{i_2} &=& (\sqrt{c_{i_2}-c_{i_1}}, & 0, & \ldots, & 0),\\ A_{i_3} &=& (\sqrt{c_{i_2}-c_{i_1}}, & \sqrt{c_{i_3}-c_{i_2}}, & \ldots, & 0),\\ \cdots & & & & & \\ A_{i_m} &=& (\sqrt{c_{i_2}-c_{i_1}}, & \sqrt{c_{i_3}-c_{i_2}}, & \ldots, & \sqrt{c_{i_m}-c_{i_{m-1}}}), \end{array}$$

clearly have the property that

$$|c_i - c_k| = |A_i - A_k|^2.$$

Thus $A \in \mathcal{S}_m$. Since the condition (5.24) is satisfied, $A \in \mathcal{P}_m$. □

Remark 5.5.12 Compare (5.25) with (4.2) for the Schlaefli simplex.

Theorem 5.5.13 *Denote by $\tilde{\mathcal{P}}_m$ the convex hull of the matrices $A = [a_{ik}]$ such that for some subset $N_0 \subset N = \{1, \ldots, m\}$, and a constant a*

$$\begin{array}{rl} a_{ik} = 0 & \text{for } \; i, k \in N_0;\\ a_{ik} = 0 & \text{for } \; i, k \in N - N_0;\\ a_{ik} = a_{ki} = a \geq 0 & \text{for } \; i \in N_0,\ k \in N - N_0. \end{array}$$

Then $\tilde{\mathcal{P}}_m \subset \mathcal{P}_m \cap \mathcal{P}_{hm} \cap \mathcal{S}_m$.

Proof. This is clear, since these matrices correspond to such systems of m points, at most two of which are distinct. □

Remark 5.5.14 The set $\tilde{\mathcal{P}}_m$ corresponds to those ordered systems of m points in $\mathcal{E}_{m-1}$, which can be completed into 2^N vertices of some right box in $\mathcal{E}_{N-1}$ (cf. Theorem 4.1.2), which may be degenerate when some opposite faces coincide.

Observe that all acute orthocentric n-simplexes also form a cone. Another, more general, observation is that the Gramians of n-simplexes also form a cone. In particular, the Gramians of the hyperacute simplexes form a cone. It is interesting that due to linearity of the expressions (5.2) and (5.3), it follows that this new operation of addition of the Gramians corresponds to addition of Menger matrices of the inverse cones.

5.6 Degenerate simplexes

Suppose that we have an n-simplex Σ and a (in a sense privileged) vector (or, direction) d. The orthogonal projection of Σ onto a hyperplane orthogonal to d forms a set of $n+1$ points in an $(n-1)$-dimensional Euclidean point space.

We can do this even as a one-parametric problem, starting with the original simplex and continuously ending with the projected object. We can then ask what happens with some distinguished points, such as the circumcenter, incenter, Lemoine point, etc. It is clear that the projection of the centroid will always exist. Thus also the vectors from the centroid to the vertices of the simplex are projected on such (linearly dependent) vectors. Forming the biorthogonal set of vectors to these, we can ask if this can be obtained by an analogous projection of some n-simplex.

We can ask what geometric object can be considered as the closest object to an n-simplex. It seems that it could be a set of $n+2$ points in the Euclidean point n-space. It is natural to assume that no $n+1$ points of these points are linearly dependent. We suggest to call such an object an *n-bisimplex*. Thus a 2-bisimplex is a quadrilateral, etc. The points which determine the bisimplex will again be called *vertices* of the bisimplex.

Theorem 5.6.1 *Let $A_1, \ldots, A_{n+2}$ be vertices of an n-bisimplex in $\mathcal{E}_n$. Then there exists a decomposition of these points into two nonvoid parts in such a way that there is a point in $\mathcal{E}_n$ which is in the convex hull of the points of each of the parts.*

Proof. The points A_i are linearly dependent but any $n+1$ of them are linearly independent. Therefore, there is exactly one (linearly independent) linear dependence relation among the points, say

$$\sum_k \alpha_k A_k = 0, \ \sum_k \alpha_k = 0;$$

here, all the α_i coefficients are different from zero. Since the sum is zero, the sets N^+ and N^- of indices corresponding to positive αs and negative αs are both nonvoid. It is then immediate that the point

$$\frac{1}{\sum_{i \in S^+} \alpha_i} \sum_{i \in S^+} \alpha_i A_i$$

coincides with the point

$$\frac{1}{\sum_{i \in S^-} \alpha_i} \sum_{i \in S^-} \alpha_i A_i$$

and has thus the mentioned property. □

Remark 5.6.2 If one of the sets S^+, S^- consists of just one element, the corresponding point is in the interior of the n-simplex determined by the remaining vertices. If one of the sets contains two elements, the corresponding points are in opposite halfspaces with respect to the hyperplane determined by the remaining vertices. Strictly speaking, only in this latter case could the result be called a bisimplex.

Suppose now that in an $(n-1)$-bisimplex in $\mathcal{E}_{n-1}$ we move one vertex in the orthogonal direction to $\mathcal{E}_{n-1}$ infinitesimally, i.e. the distance of the moved vertex from $\mathcal{E}_{n-1}$ will be $\epsilon > 0$ and tending to zero. The resulting object will be called a *degenerate n-simplex.* Every interior angle of this n-simplex will be either infinitesimally small, or infinitesimally close to π.

Example 5.6.3 Let A, B, C, D be the points from Example 5.5.1. If the point D has the third coordinate ϵ and the remaining three zero, the resulting tetrahedron will have four acute angles opposite the edges AD, DC, AB, and BC, and two obtuse angles opposite the edges AC and BD.

We leave it to the reader to show that a general theorem on the colored graph (cf. Chapter 3, Section 1) of the degenerate n-simplex holds.

Theorem 5.6.4 *Let $A_1, \ldots, A_{n+1}$ be vertices of a degenerate n-simplex, and let N^+, N^- be the parts of the decomposition of indices as in the proof of Theorem 5.6.1. Then the edges A_iA_k will be red if and only if i and k belong to different sets N^+, N^-, and blue if and only if i and k belong to the same set N^+, or N^-.*

Remark 5.6.5 Theorem 5.6.4 implies that the degenerate simplex is flat in the sense of Remark 2.2.20. In fact, this was the reason for that notation.

Another type of an n-simplex close to degenerate is one which we suggest be called a *needle.* It is essentially a simplex obtained by perturbations from the set of $n+1$ points on a line. Such a simplex should have the property that every face is again a needle and, if possible, the colored graph of every face (as well as of the simplex itself) should have a red path containing all the vertices, whereas all the remaining edges are blue. One possibility is using the following result. First, call a tetrahedron $A_1A_2A_3A_4$ with *ordered* vertices a *t-needle* if the angles $\angle A_1A_2A_3$ and $\angle A_2A_3A_4$ are obtuse and the sum of squares $|A_1A_3|^2 + |A_2A_4|^2$ is *smaller* than $|A_1A_4|^2 + |A_2A_3|^2$. Then if each tetrahedron $A_iA_{i+1}A_{i+2}A_{i+3}$ for $i = 1, \ldots, n-2$ is a t-needle, then *all* the tetrahedrons $A_{i_1}A_{i_2}A_{i_3}A_{i_4}$ are t-needles if $i_1 < i_2 < i_3 < i_4$.

6
Applications

6.1 An application to graph theory

In this section, we shall investigate undirected graphs with n nodes without loops and multiple edges. We assume that the set of nodes is $N = \{1, 2, \dots, n\}$ and we write $G = (N, E)$ where E denotes the set of edges.

Recall that the *Laplacian matrix* of $G = (N, E)$, *Laplacian* for short, is the real symmetric $n \times n$ matrix $L(G)$ whose quadratic form is

$$\langle L(G)x, x\rangle = \sum_{i,k,i<k,(i,k)\in E} (x_i - x_k)^2.$$

Let us list a few elementary properties of $L(G)$:

(L 1) $L(G) = D(G) - A(G)$,
where $A(G)$ is the adjacency matrix of G and $D(G)$ is the diagonal matrix, the ith diagonal entry of which is d_i, the degree of the node i in G;
(L 2) $L(G)$ is positive semidefinite and singular;
(L 3) $L(G)e = 0$, where e is the vector of all ones;
(L 4) if G is connected, then $L(G)$ has rank $n - 1$;
(L 5) for the complement $\bar{G}$ of G, $L(G) + L(\bar{G}) = nI - J$,
where $J = ee^T$.

More generally, we can define the Laplacian of a weighted graph $G_C = (N, E, C)$ with nonnegative weight $c_{ij} = c_{ji}$ assigned to each edge $(i, j) \in E$ as follows:

$L(G_C)$ is the symmetric matrix of the quadratic form

$$\sum_{(i,j)\in E,i<j} c_{ij}(x_i - x_j)^2.$$

Clearly, $L(G_C)$ is again positive semidefinite and singular with $L(G_C)e = 0$. If the graph with the node set N and the set of positively weighted edges in C is connected, the rank of $L(G_C)$ is $n-1$. In this last case, we shall call such a weighted graph a *connected weighted graph.*

Now let G (or G_C) be a (weighted) graph with n nodes. The eigenvalues

$$\lambda_1 \leq \lambda_2 \leq \ldots \leq \lambda_n$$

of $L(G)$ (or $L(G_C)$) will be called *Laplacian eigenvalues* of G (G_C) (λ_1 the first, λ_2 the second etc.).

The first (smallest) Laplacian eigenvalue λ_1 is zero, the second λ_2 was denoted as $a(G)$ and called the *algebraic connectivity* of the graph G in [4] since it has similar properties as the edge-connectivity $e(G)$, the number of edges (or, the sum of edge-weights) in the minimum cut.

In addition, the following inequalities were proved in [17]

$$2\Big(1 - \cos\frac{\pi}{n}\Big)e(G) \leq a(G) \leq e(G).$$

An important property of the eigenvector corresponding to λ_2 was proved in [19].

Theorem 6.1.1 *Let G be a connected graph, and let u be a (real) eigenvector of $L(G)$ corresponding to the algebraic connectivity $a(G)$. Then the subgraph of G induced by the node set corresponding to the nodes with nonnegative coordinates of u is connected.*

We shall first prove the following lemma.

Lemma 6.1.2 *Suppose A is an $n \times n$ symmetric nonnegative irreducible matrix with eigenvalues $\lambda_1 \geq \lambda_2 \geq \cdots \geq \lambda_n$. Denote by $v = [v_i]$ a (real) eigenvector of A corresponding to λ_2 and $N = \{1, 2, \ldots, n\}$. If $M = \{i \in N; v_i \geq 0\}$, then M is a proper nonvoid subset of N and the principal submatrix $A(M)$ with rows and columns indices in M is irreducible.*

Proof. Without loss of generality, we can choose $M = \{1, 2, \ldots, m\}$. Suppose that $A(M)$ is reducible and of the form

$$A(M) = \begin{bmatrix} A_{11} & & & \\ & A_{22} & & \\ & & \ddots & \\ & & & A_{rr} \end{bmatrix}, \quad r \geq 2,$$

where all A_{ii}s are irreducible. Thus A has the form

$$A = \begin{bmatrix} A_{11} & & 0 & A_{1,r+1} \\ & \ddots & & \vdots \\ 0 & & A_{rr} & A_{r,r+1} \\ A_{1,r+1}^T & \cdots & A_{r,r+1}^T & A_{r+1,r+1} \end{bmatrix}.$$

Let

$$v = \begin{bmatrix} v^{(1)} \\ \vdots \\ v^{(r)} \\ v^{(r+1)} \end{bmatrix}$$

be the corresponding partitioning of v, $v^{(1)} \geq 0, \ldots, v^{(r)} \geq 0$, $v^{(r+1)} < 0$. Then

$$(A_{kk} - \lambda_2 I_k)v^{(k)} = -A_{k,r+1}v^{(r+1)}, \quad k = 1, \ldots, r. \tag{6.1}$$

The matrix $\lambda_2 I - A$ has (since $\lambda_1 > \lambda_2$ by the Perron–Frobenius theorem) exactly one negative eigenvalue; therefore, its principal submatrix $\lambda_2 I(M) - A(M)$ has at most one negative eigenvalue. Since $r \geq 2$, some diagonal block, say $\lambda_k I_k - A_{kk}$, has only nonnegative eigenvalues and is thus a (nonsingular or singular) M-matrix. If it were singular, then by Theorem A.3.9 we would have

$$(\lambda_2 I_k - A_{kk})z^{(k)} = 0 \quad \text{for } z^{(k)} > 0.$$

Multiplication of (6.1) by $(z^{(k)})^T$ from the left implies

$$(z^{(k)})^T A_{k,r+1} v^{(r+1)} = 0,$$

i.e. $A_{k,r+1} = 0$, a contradiction of irreducibility.

Therefore, $\lambda_k I_k - A_{kk}$ is nonsingular and, by the property of M-matrices, $(\lambda_k I_k - A_{kk})^{-1} > 0$. Now, (6.1) implies

$$v^{(k)} = (\lambda_k I_k - A_{kk})^{-1} A_{k,r+1} v^{(r+1)};$$

the left-hand side is a nonnegative vector, the right-hand side a nonpositive vector. Consequently, both are equal to zero, and necessarily $A_{k,r+1} = 0$, a contradiction of irreducibility again.

Let us complete the proof of Theorem 6.1.1. Choose a real c so that the matrix $A = cI - L(G)$ is nonnegative. The maximal eigenvalue of A is c, and λ_2 is $c - a(G)$, with the corresponding eigenvector v a real multiple of u. By the lemma, if this multiple is positive, the subgraph of G induced by the set of vertices with nonnegative coordinates is connected. Since the lemma also holds for the vector $-v$, the result does not depend on multiplication by -1. $\square$

6.2 Simplex of a graph

In this section, we assume that G is a connected graph. Observe that the Laplacian $L(G)$ as well as the Laplacian $L(G_C)$ of a connected weighted graph G_C satisfy the conditions (1.31) and (1.32) (with n instead of $n+1$) of the matrix Q.

Therefore, we can assign to G [or G_C] in $\mathcal{E}_{n-1}$ an – up to congruence uniquely defined – $(n-1)$-simplex $\Sigma(G)$ [or, $\Sigma(G_C)$], which we shall call the *simplex of the graph* G [G_C, respectively]. The corresponding Menger matrix M will be called the *Menger matrix* of G.

Applying Theorem 1.4.1, we immediately have the following theorem, formulated just for the more general case of a weighted graph:

Theorem 6.2.1 *Let e be the column vector of n ones. Then there exists a unique column vector q_0 and a unique number q_{00} such that the symmetric matrix $M = [m_{ik}]$ with $m_{ii} = 0$ satisfies*

$$\begin{bmatrix} 0 & e^T \\ e & M \end{bmatrix} \begin{bmatrix} q_{00} & q_0^T \\ q_0 & L(G_C) \end{bmatrix} = -2I_{n+1}. \tag{6.2}$$

The (unique) matrix M is the Menger matrix of G_C.

Example 6.2.2 Let G be the path P_4 with four nodes 1, 2, 3, 4 and edges $(1,2)$, $(2,3)$, $(3,4)$. Then (6.2) reads as follows

$$\underbrace{\left[\begin{array}{c|cccc} 0 & 1 & 1 & 1 & 1 \\ \hline 1 & 0 & 1 & 2 & 3 \\ 1 & 1 & 0 & 1 & 2 \\ 1 & 2 & 1 & 0 & 1 \\ 1 & 3 & 2 & 1 & 0 \end{array}\right]}_{M} \underbrace{\left[\begin{array}{c|cccc} 3 & -1 & 0 & 0 & -1 \\ \hline -1 & 1 & -1 & 0 & 0 \\ 0 & -1 & 2 & -1 & 0 \\ 0 & 0 & -1 & 2 & -1 \\ -1 & 0 & 0 & -1 & 1 \end{array}\right]}_{L(P_4)} = -2I_5.$$

Example 6.2.3 For a star S_n with nodes $1, 2, \ldots, n$ and the set of edges $(1,k)$, $k = 2, \ldots, n$, the equality (6.2) reads

$$\begin{bmatrix} 0 & 1 & 1 & 1 & \ldots & 1 \\ 1 & 0 & 1 & 1 & \ldots & 1 \\ 1 & 1 & 0 & 2 & \ldots & 2 \\ 1 & 1 & 2 & 0 & \ldots & 2 \\ \cdot & \cdot & \cdot & \cdot & \cdot & \cdot \\ 1 & 1 & 2 & 2 & \ldots & 0 \end{bmatrix} \begin{bmatrix} n-1 & n-3 & -1 & -1 & \ldots & -1 \\ n-3 & n-1 & -1 & -1 & \ldots & -1 \\ -1 & -1 & 1 & 0 & \ldots & 0 \\ -1 & -1 & 0 & 1 & \ldots & 0 \\ \cdot & \cdot & \cdot & \cdot & \cdot & \cdot \\ -1 & -1 & 0 & 0 & \ldots & 1 \end{bmatrix}$$
$$= -2I_{n+1}.$$

Remark 6.2.4 Observe that – in agreement with Theorem 4.1.3 – in both cases, the Menger matrix M is at the same time the *distance* matrix of G, i.e. the matrix $D = [D_{ik}]$ for which D_{ik} means the distance between the nodes i and k in G_C, in general the minimum of the lengths of all the paths between i and k, the length of a path being the sum of the lengths of edges contained in the path. We intend to prove that $M = D$ for all weighted trees, of which the length of each edge is appropriately chosen.

Theorem 6.2.5 *Let $T_C = (N, E, C)$ be a connected weighted tree, $N = \{1, \dots, n\}$, and $c_{ik} = c_{ki}$ denote the weight of the edge $(i,k) \in E$. Let $L(T_C) = [q_{ik}]$ be the Laplacian of T_C so that for $i, k \in N$*

$$\begin{aligned} q_{ii} &= \sum_{k,(i,k)\in E} c_{ik}, \\ q_{ik} &= -c_{ik} \text{ for } (i,k) \in E, \\ q_{ik} &= 0 \text{ otherwise.} \end{aligned}$$

Further denote for $i, j, k \in N$

$$\begin{aligned} R_{ij} &= \frac{1}{c_{ij}} \text{ for } (i,j) \in E, \\ R_{ii} &= 0, \\ R_{ik}(= R_{ki}) &= R_{ij_1} + R_{j_1 j_2} + \cdots + R_{j_{s-1},j_s} + R_{j_s,k} \end{aligned}$$

whenever $(i, j_1, j_2, \dots, j_s, k)$ is the (unique) path from i to k in T_C.

If, moreover

$$\begin{aligned} R_{00} &= 0, \\ R_{0i} &= 1,\ i \in N, \\ q_{00} &= \sum_{(i,k)\in E, i<k} R_{ik}, \\ q_{0i} &= d_i - 2,\ d_i \text{ being the degree of } i \in N \text{ in } T_C, \end{aligned}$$

then the symmetric matrices

$$\widetilde{R} = [R_{rs}], \quad \widetilde{Q} = [q_{rs}], \quad r, s = 0, 1, \dots, n$$

satisfy

$$\widetilde{R}\widetilde{Q} = -2I_{n+1}.$$

Proof. We apply Lemma 4.1.1 with R_{rs} instead of m_{rs} and q_{rs} instead of γ_{rs} for $r, s = 0, 1, \dots, n+1$. The result follows. □

For $c_{ik} = 1$ for all $i,\ k$, we obtain:

Corollary 6.2.6 *Let $T = (N, E)$ be a tree with n nodes. Denote by $z = [z_i]$ the column n-vector with $z_i = d_i - 2$, d_i being the degree of the node i. Then:*

(i) *the Menger matrix M of T coincides with the (usual) distance matrix D of T;*
(ii) *the equality (6.2) reads*

$$\begin{bmatrix} 0 & e^T \\ e & D \end{bmatrix} \begin{bmatrix} n-1 & z^T \\ z & L(T) \end{bmatrix} = -2I_{n+1}.$$

Let us return now to Theorem 5.1.2 and formulate a result which will relate the Laplacian eigenvalues to the Menger matrix M.

Theorem 6.2.7 *Let M be the Menger matrix of a graph G. Then the $n-1$ roots of the equation*

$$\det \begin{bmatrix} 0 & e^T \\ e & M - \mu I \end{bmatrix} = 0 \tag{6.3}$$

are the numbers $\mu_i = -2/\lambda_i$, $i = 2, \ldots, n$ where $\lambda_2, \ldots, \lambda_n$ are the nonzero Laplacian eigenvalues of G. If $y^{(i)}$ is an eigenvector of $L(G)$ corresponding to $\lambda_i \neq 0$, then $\begin{bmatrix} \lambda_0^{-1} q_0^T y^{(i)} \\ y^{(i)} \end{bmatrix}$ is the corresponding annihilating vector of the matrix in (6.3).

Proof. The first part follows immediately from the identity

$$\begin{bmatrix} 0 & e^T \\ e & M - \mu I \end{bmatrix} \begin{bmatrix} q_{00} & q_0^T \\ q_0 & L(G) \end{bmatrix} = \begin{bmatrix} -2 & 0 \\ * & -2I - \mu L(G) \end{bmatrix}, \tag{6.4}$$

where $*$ is some column vector. Postmultiplying (6.4) with $\mu = -2/\lambda_i$ by $\begin{bmatrix} 0 \\ y^{(i)} \end{bmatrix}$, we obtain the second assertion. □

Corollary 6.2.8 *Let G be connected and $L(G) = ZZ^T$ be any full-rank-factorization, so that Z is an $n \times (n-1)$ matrix. Then*

$$Z^T M Z = -2I_{n-1}, \tag{6.5}$$

where M is the Menger matrix of G.

Proof. By (6.2)

$$ML(G) = -2I_n - eq_0^T.$$

Since $Z^T e = 0$, by premultiplying by Z^T we get

$$Z^T M Z Z^T = -2Z^T,$$

and hence (6.5) holds. □

The interlacing theorem (cf. Appendix, Theorem A.1.33) will now be applied.

Corollary 6.2.9 *The roots of (6.3), i.e. the numbers $-2/\lambda_i$ where the λ_i are the nonzero Laplacian eigenvalues of G, $i = 2, \ldots, n$, interlace the eigenvalues of the Menger matrix M of G.*

Proof. This follows immediately from the following Lemma which is easily proved by transforming A to diagonal form by an orthogonal transformation: □

Lemma 6.2.10 *Let A be a real symmetric matrix, u be a nonzero real vector, and t be a real number. Then the zeros of the equation*

$$\det \begin{bmatrix} t & u^t \\ u & A - xI \end{bmatrix} = 0$$

interlace the eigenvalues of A.

Using the previous Lemma, an analogous proof to that of Theorem 6.2.7 gives:

Theorem 6.2.11 *The eigenvalues m_i of the Menger matrix M of G and the roots x_i of the equation*

$$\det \begin{bmatrix} q_{00} & q_0^T \\ q_0 & L(G) - xI \end{bmatrix} = 0$$

satisfy

$$m_i x_i = -2, \quad i = 1, \ldots, n.$$

The numbers x_i and the Laplacian eigenvalues λ_i of G interlace each other.

Let us return now to the geometric considerations from Section 4.

Theorem 6.2.12 *Let λ_i be a nonzero eigenvalue of $L(G_C)$, and y be a corresponding eigenvector. Then*

$$\frac{n-1}{n} \cdot \frac{1}{\lambda_i}$$

is the square of a halfaxis of the Steiner circumscribed ellipsoid of the simplex $\Sigma(G_C)$ of G_C and $\Sigma y_i x_i = 0$ is the equation of the hyperplane orthogonal to the corresponding axis. Also, y is the direction of the axis.

Corollary 6.2.13 *The smallest positive eigenvalue of $L(G_C)$ (the algebraic connectivity of G_C) corresponds to the* largest *halfaxis of the Steiner circumscribed ellipsoid of $\Sigma(G_C)$.*

Due to the result presented here as Theorem 6.1.1, the eigenvector $y = [y_i]$ corresponding to the second smallest eigenvalue $a(G)$ of G seems to be a good separator in the set of nodes N of G in the sense that the ratio of the cardinalities of the two parts is neither very small nor very large. The geometric meaning of the subsets N^+ and N^- of N is as follows.

Theorem 6.2.14 *Let $y = [y_i]$ be an eigenvector of $L(G)$ corresponding to λ_2 $(= a(G))$. Then*

$$\begin{aligned} N^+ &= \{i \in N \mid y_i > 0\}, \\ N^- &= \{i \in N \mid y_i < 0\}, \\ Z &= \{i \in N \mid y_i = 0\} \end{aligned}$$

correspond to the number of vertices of the simplex $\Sigma(G)$ in the decomposition of $\mathcal{E}_n$ with respect to the hyperplane of symmetry H of the Steiner circumscribed ellipsoid orthogonal to the largest halfaxis: $|N^+|$ is the number of vertices of $\Sigma(G)$ in one halfspace H^+, $|N^-|$ is the number of vertices in H^-, and $|Z|$ is the number of vertices in H.

6.3 Geometric inequalities

We add here a short section on inequalities, which was prompted from the previous considerations.

Theorem 6.3.1 *Let Σ be an n-simplex, with F_1 and F_2 its $(n-1)$-dimensional faces. Then their volumes satisfy*

$$n\mathrm{V}_n(\Sigma)\mathrm{V}_{n-2}(F_1 \cap F_2) \leq \mathrm{V}_{n-1}(F_1)\mathrm{V}_{n-1}(F_2).$$

Equality is attained if and only if the faces F_1 and F_2 are orthogonal.

Proof. Follows from the formula (2.4). □

Let us show how this formula can be generalized.

Theorem 6.3.2 *Let Σ be an n-simplex, with F_1 and F_2 its faces with nonvoid intersection. Then the volumes of the faces $F_1 \cap F_2$ and $F_1 \cup F_2$ (the smallest face containing both F_1 and F_2) satisfy: if f_1, f_2, f_0, and f are the dimensions of $F_1, F_2, F_1 \cap F_2$, and $F_1 \cup F_2$, respectively, then*

$$f!f_0!\mathrm{V}_f(F_1 \cup F_2)\mathrm{V}_{f_0}(F_1 \cap F_2) \leq f_1!f_2!\mathrm{V}_{f_1}(F_1)\mathrm{V}_{f_2}(F_2). \tag{6.6}$$

Equality is attained if and only if the faces F_1 and F_2 are orthogonal in the space of $F_1 \cup F_2$.

Proof. Suppose that the vertex A_{n+1} is in the intersection. The $n \times n$ matrix $\hat{M}$ with (i, j) entries $m_{i,n+1} + m_{j,n+1} - m_{ij}$ is then positive definite and each of its principal minors has determinant corresponding to the volume of a face. Using now the Hadamard–Fischer inequality (Appendix, (A.13)), we obtain (6.6). The proof of equality follows from considering the Schur complements and the case of equality in (A.12). □

An interesting inequality for the altitudes of a spherical simplex was proved in Theorem 5.4.5. We can use it also for the usual n-simplex for the limiting case when the radius grows to infinity. We have already proved the strict polygonal inequality for the reciprocals of the lengths l_is in (iii) of Theorem 2.1.4 using the volumes of the $(n-1)$-dimensional faces

$$2 \max_i \frac{1}{l_i} < \sum_i \frac{1}{l_i}.$$

By (2.3), this generalizes the triangle inequality.

In Example 2.1.11, we obtained the formula

$$576r^2V^2 = (aa' + bb' + cc')(-aa' + bb' + cc')(aa' - bb' + cc')(aa' + bb' - cc').$$

Since the left-hand side is positive, it follows that all expressions in the parentheses on the right-hand side have to be positive. Thus

$$2\max(aa', bb', cc') \leq aa' + bb' + cc'.$$

This is a generalization of the *Ptolemy inequality*: The products of the lengths of two opposite pairs among four points do not exceed the sum of the products for the remaining pairs. We proved it for the three-dimensional space, but a limiting procedure leads to the more usual plane case as well.

Many geometric inequalities follow from solutions of optimization problems. Usually, the regular simplex is optimal. Let us present an example.

Theorem 6.3.3 *If the circumcenter of an n-simplex Σ is an interior point of Σ or a point of its boundary, then the length of the maximum edge of Σ is at least $r\sqrt{2(n+1)/n}$, where r is the circumradius. Equality is attained for the regular n-simplex.*

Proof. Let C be the circumcenter in Σ with vertices A_i. Denote $u_i = A_i - C$, $i = 1, \ldots, n+1$. The condition about the position of C means that all the coefficients μ_i in the linear dependence relation $\sum_i \mu_i u_i = 0$ are nonnegative. We have then $\langle u_i, u_i \rangle = r^2$: supposing that $|A_iA_j|^2 < 2(n+1)n^{-1}r^2$ or, $\langle u_i - u_j, u_i - u_j \rangle < (n+1)n^{-1}r^2$, for every pair i, j, $i \neq j$, means that for each such pair $\langle u_i, u_j \rangle > -1/n$. Thus $\left\langle \sum_i \mu_i u_i, u_k \right\rangle = 0$ for each k implies after dividing by r^2 that

$$0 > \mu_k - \frac{1}{n} \sum_{i \neq k} \mu_i$$

for each k. But summation over all k yields $0 > 0$, a contradiction. □

6.4 Extended graphs of tetrahedrons

Theorems in previous chapters can be used for finding all possible extended graphs of tetrahedrons, i.e. extended graphs of simplexes with five nodes. Most general theorems were proved already in Section 3.4. We start with a lemma.

Lemma 6.4.1 *Suppose that G_0 is an extended graph of a tetrahedron with the nodes $0, 1, 2, 3, 4$, and G its induced subgraph with nodes $1, 2, 3, 4$. If G has exactly one negative edge (i, j), then $(0, i)$ as well as $(0, j)$ are positive edges in G_0.*

Proof. Without loss of generality, assume that $(1,2)$ is that negative edge in G. Let Σ be the corresponding tetrahedron with graph G and extended graph G_0. The Gramian $Q = [q_{ik}]$, $i,k = 1,2,3,4$, is then positive semidefinite with rank 3, and satisfies $Qe = 0$, and $q_{12} > 0$, $q_{ij} \leq 0$ for all the remaining pairs i,j, $i \neq j$.

Since $q_{11} > 0$, we have $q_{12} + q_{13} + q_{14} < 0$. By (3.9)

$$\begin{aligned} c_1 = & -(-q_{12})(-q_{13})(-q_{14}) + (-q_{12})(-q_{23})(-q_{24}) + \\ & +(-q_{13})(-q_{23})(-q_{34}) + (-q_{14})(-q_{24})(-q_{34}) + \\ & +(-q_{12})(-q_{23})(-q_{34}) + (-q_{12})(-q_{24})(-q_{34}) + \\ & +(-q_{13})(-q_{34})(-q_{24}) + (-q_{13})(-q_{23})(-q_{24}) + \\ & +(-q_{14})(-q_{24})(-q_{23}) + (-q_{14})(-q_{34})(-q_{23}), \end{aligned}$$

since at the summands corresponding to the remaining spanning trees in G the coefficient $2 - k_i$ is zero. Thus

$$c_1 = q_{12}q_{13}q_{14} - (q_{12} + q_{13} + q_{14})(q_{23}q_{24} + q_{23}q_{34} + q_{24}q_{34}) > 0,$$

since both summands are nonnegative and at least one is positive, the positive part of G being connected by Theorem 3.4.6. Therefore, the edge $(0,1)$ is positive in G_0. The same holds for the edge $(0,2)$ by exchanging the indices 1 and 2. □

We are able now to prove the main theorem on extended graphs of tetrahedrons.

Theorem 6.4.2 *None of the graphs P_5, Q_5, or R_5 in Fig. 6.1 as well as none of its subgraphs with five nodes, with the exception of the positive circuit, can be an extended graph of a tetrahedron. All the remaining signed graphs on five nodes, the positive part of which has edge-connectivity at least two, can serve as extended graphs of tetrahedrons (with an arbitrary choice of the vertex corresponding to the circumcenter). There are* 20 *such (mutually non-isomorphic) graphs. They are depicted in Fig. 6.2.*

Proof. Theorem 3.4.21 shows that neither P_5, Q_5, R_5, nor any of their subgraphs on five vertices with at least one negative edge, can be extended graphs of a tetrahedron. Since the positive parts of P_5, Q_5, R_5 have edge-connectivity at least two, the assertion in the first part holds by Theorem 3.4.15.

Fig. 6.1

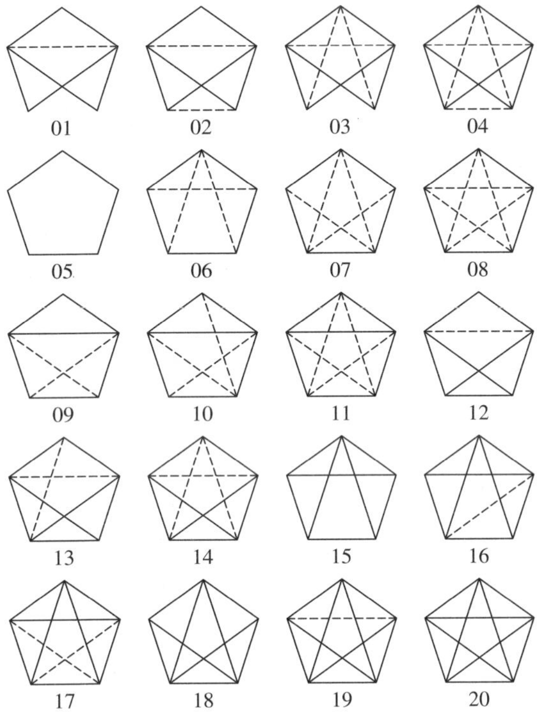

Fig. 6.2

To prove the second part, construct in Fig. 6.2 all those signed graphs on five nodes, the positive part of which has edge-connectivity at least two. From the fact that every positive graph on five nodes with edge-connectivity at least two must contain the graph 05 or the positive part of 01, it follows that the list is complete. Let us show that all these graphs are extended graphs of tetrahedrons. We show this first for graph 09. The graph in Fig. 6.3 is the (usual) graph of some tetrahedron by Theorem 3.1.3. The extended graph of this tetrahedron contains by (ii) of Theorem 3.4.10 the positive edge $(0, 4)$, and by Lemma 6.4.1 positive edges $(0, 1)$ and $(0, 3)$. Assume that $(0, 2)$ is either positive or missing. In the first case the edge $(0, 3)$ would be positive by (iv) of Theorem 3.4.10 (if we remove node 3); in the second, $(0, 3)$ would be missing by (i) of Theorem 3.4.10. This contradiction shows that $(0, 2)$ is negative and the graph 09 is an extended graph of a tetrahedron. The graphs 01 and 05 are extended graphs of right simplexes by (iv) of Theorem 3.4.10. To prove that the graphs 06 and 15 are such extended graphs, we need the assertion that the graph of Fig. 6.4 is the (usual) graph of the obtuse cyclic tetrahedron (a special case of the cyclic simplex from Chapter 4, Section 3). By Lemma 6.4.1, the extended graph has to contain positive edges $(0, 1)$ and

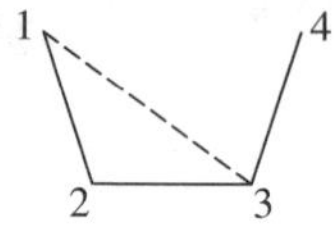

Fig. 6.3

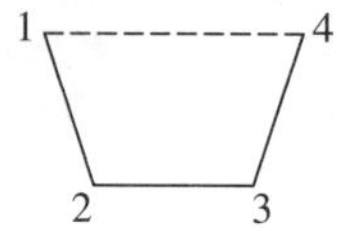

Fig. 6.4

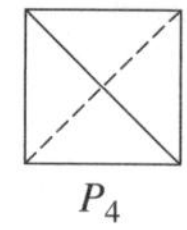

Fig. 6.5

$(0, 4)$. Using the formulae (3.9) for c_2 and c_3, we obtain that $c_2 < 0$ and $c_3 < 0$. Thus 06 is an extended graph. For the graph 15, the proof is analogous. By Theorem 3.4.21, the following graphs are such extended graphs: 2, 3, 4, 12, 13, 14 (they contain the graph 01), 7, 8 (contain the graph 06), 10, 11 (contain the graph 09), 16, 17, 18, 19, and 20 (contain the graph 15). The proof is complete. □

In a similar way as in Theorem 6.4.2, the characterization of extended graphs of triangles can be formulated:

Theorem 6.4.3 *Neither the graph P_4 from Fig. 6.5, nor any of its subgraphs on four nodes, with the exception of the positive circuit, can be the extended graph of a triangle. All the remaining signed graphs with four nodes, whose positive part has edge-connectivity at least two, are extended graphs of a triangle.*

Proof. Follows immediately by comparing the possibilities with the graphs in Fig. 3.2 in Section 3.4. □

Remark 6.4.4 As we already noticed, the problem of characterizing all extended graphs of n-simplexes is for $n > 3$ open.

6.5 Resistive electrical networks

In this section, we intend to show an application of geometric and graph-theoretic notions of the previous chapters to resistive electrical networks.

In the papers [20], [26] the author showed that if $n \geq 2$, the following four sets are mathematically equivalent:

A. $\mathcal{G}_n$, the set of (in a sense connected) nonnegative valuations of a complete graph with n nodes $1, \ldots, n$;

B. $\mathcal{M}_n$, the set of real symmetric $n \times n$ M-matrices of rank $n-1$ with row-sums zero;

C. $\mathcal{N}_n$, the set of all connected electrical networks with n nodes and such that the branches contain resistors only;

D. $\mathcal{S}_n$, the set of all (in fact, classes of mutually congruent) hyperacute $(n-1)$-simplexes with numbered vertices.

The parameters which in separate cases determine the situations are:

(A) nonnegative weights $w_{ik} (= w_{ki})$, $i, k = 1, \ldots, n$, $i \neq k$ assigned to the edges (i, k);
(B) the negatives of the entries $a_{ik} = a_{ki}$ of the matrix $[a_{ik}]$, $i \neq k$;
(C) the conductivities C_{ik} (the inverses of resistances) in the branch between i and k, if it is contained in the network, or zero, if not;
(D) the negatives of the entries $q_{ik} = q_{ki}$ of the Gramian of the $(n-1)$-simplex (or, the class).

Observe that the matrix of the graph in (B) is the Laplacian of the graph in (A) and that the $(n-1)$-simplex in (D) is the simplex of the graph in the sense of Section 6.1.

The equivalent models have then the determining elements identical. The advantage is, of course, that we can use methods of each model in the other models, interpret them, etc. In addition, there are some common "invariants."

Theorem 6.5.1 *The following values are in each model the same:*
(A) edge-connectivity $e(G)$*;*
(B) the so-called measure of irreducibility of the matrix A*, i.e.* $\min_{M, \emptyset \neq M \neq N} \sum_{i \in M, k \notin M} |a_{ik}|$*;*
(C) the minimal conductivity between complementary groups of nodes;
(D) the reciprocal value of the maximum of the squares of distances between pairs of complementary faces of the simplex, in other words the reciprocal of the square of the thickness of the simplex.

Probably most interesting is the following equivalence (in particular between (C) and (D)):

Theorem 6.5.2 *Let* i *and* j *be distinct indices. Then the following quantities are equivalent:*
(A) $\sum_{H \in K_{ij}} \pi(H) / \sum_{H \in K} \pi(H)$*, where for the subgraph* H *of the weighted graph* G*,* $\pi(H)$ *means the sum of all weights on the edges of* H*,* K *is the set of spanning trees of* G*, and* K_{ij} *is the set of all forests in* G *with two components: one containing the node* i*, the second the node* j*.*
(B) $\det A(N \backslash \{i, j\}) / \det A(N \backslash \{i\})$*, where* $N = \{1, \ldots, n\}$ *and* $A(M)$ *for* $M \subset N$ *is the principal submatrix of the matrix* A *with rows in* M*.*
(C) R_{ij}*, which is the global resistance in the network between the* i*th and* j*th node.*
(D) m_{ij}*, i.e. the square of the distance between the* i*th and* j*th vertex of the simplex.*

The proof is in [20]. It is interesting that the equivalence between (C) and (D), which the author knew in 1962, was discovered and published in 1968 by D. J. H. Moore [32].

We conclude the section by a comment on a closer relationship between the models (C) and (D) (although it could be of interest to consider the corresponding objects in the models (A) and (B) as well).

Let us observe first that the equivalence of (C) and (D) in Theorem 6.5.2 answers the question about all possible structures of mutual resistances, which can occur between the n outlets of a black box if we know that the net is connected and contains resistors only. It is identical to the structure of the squares of distances in hyperacute $(n-1)$-simplexes, i.e. of their Menger matrices.

Furthermore, the theorem that every at least two-dimensional face of a hyperacute simplex is also hyperacute implies that the smaller black box obtained from a black box by ignoring some of its outlets is also a realizable black box itself.

It also follows that every realizable resistive black box can be realized by a complete network in which every outlet is connected with every other outlet by some resistive branch. Conversely we can possibly use this equivalence for finding networks with the smallest structure (by adding further auxiliary nodes).

We can also investigate what happens if we make shortcuts between two or more nodes in such a network. This corresponds to an orthogonal projection of the simplex along the face connecting the shortcut vertices. Geometrically, this means that every such projection is again a hyperacute simplex.

From the network theory it is known that if we put on two disjoint sets of outlets a potential (that means that we join the nodes of each of these sets by a shortcut and then put the potential between), then each of the remaining nodes will have some potential. Geometrically this potential can be found as follows: we find the layer of the maximum thickness which contains on one of the boundary hyperplanes the vertices of the first group and on the other those of the second group. Thanks to the hyperacuteness property, all vertices of the simplex are contained in the layer. The ratio of the distances to the first and the second hyperplane determines then the potential of each remaining vertex. Of course, the maximum layer is obtained by such position of the boundary hyperplane in which it is orthogonal to the linear space determined by the union of both groups.

It would be desirable to find the interpretation of the numbers q_{0i} and q_{00} in the electrical model and of q_{00} in the graph-theoretical model.

Appendix

A.1 Matrices

Throughout the book, we use basic facts from matrix theory and the theory of determinants. The interested reader may find the omitted proofs in general matrix theory books, such as [28], [31], and others.

A *matrix of type m-by-n* or, equivalently, an $m \times n$ matrix, is a two-dimensional array of mn numbers (usually real or complex) arranged in m *rows* and n *columns* (m, n positive integers)

$$\begin{bmatrix} a_{11} & a_{12} & a_{13} & \dots & a_{1n} \\ a_{21} & a_{22} & a_{23} & \dots & a_{2n} \\ \cdot & \cdot & \cdot & \dots & \cdot \\ a_{m1} & a_{m2} & a_{m3} & \dots & a_{mn} \end{bmatrix}. \tag{A.1}$$

We call the number a_{ik} the *entry* of the matrix (A.1) in the ith row and the kth column. It is advantageous to denote the matrix (A.1) by a single symbol, say A, C, etc. The set of $m \times n$ matrices with real entries is denoted by $R^{m \times n}$. In some cases, $m \times n$ matrices with complex entries will occur and their set is denoted analogously by $C^{m \times n}$. In some cases, entries can be polynomials, variables, functions, etc.

In this terminology, matrices with only one column (thus, $n = 1$) are called *column vectors*, and matrices with only one row (thus, $m = 1$) *row vectors.* In such a case, we write R^m instead of $R^{m \times 1}$ and – unless said otherwise – vectors are always column vectors.

Matrices of the same type can be added entrywise: if $A = [a_{ik}]$, $B = [b_{ik}]$, then $A+B$ is the matrix $[a_{ik}+b_{ik}]$. We also admit multiplication of a matrix by a number (real, complex, a parameter, etc.). If $A = [a_{ik}]$ and if α is a number (also called *scalar*), then αA is the matrix $[\alpha a_{ik}]$, of the same type as A.

An $m \times n$ matrix $A = [a_{ik}]$ can be multiplied by an $n \times p$ matrix $B = [b_{kl}]$ as follows: AB is the $m \times p$ matrix $C = [c_{il}]$, where

$$c_{il} = a_{i1}b_{1l} + a_{i2}b_{2l} + \cdots + a_{in}b_{nl}.$$

It is important to notice that the matrices A and B can be multiplied (in this order) only if the number of columns of A is the same as the number of rows in B. Also, the entries of A and B should be multiplicable. In general, the product AB is not equal to BA, even if the multiplication of both products is possible. On the other hand, the multiplication fulfills the *associative law*

$$(AB)C = A(BC)$$

as well as (in this case, two) *distributive laws*:

$$(A + B)C = AC + BC$$

and

$$A(B + C) = AB + AC,$$

whenever multiplications are possible.

Of basic importance are the *zero matrices*, all entries of which are zeros, and the *identity matrices*; the latter are *square* matrices, i.e. $m = n$, and have ones in the *main diagonal* and zeros elsewhere. Thus

$$[1], \quad \begin{bmatrix} 1 & 0 \\ 0 & 1 \end{bmatrix}, \quad \begin{bmatrix} 1 & 0 & 0 \\ 0 & 1 & 0 \\ 0 & 0 & 1 \end{bmatrix}$$

are identity matrices of *order* one, two, and three. We denote the zero matrices simply by 0, and the identity matrices by I, sometimes with a subscript denoting the order.

The identity matrices of appropriate orders have the property that

$$AI = A \text{ and } IA = A$$

hold for any matrix A.

Now let $A = [a_{ik}]$ be an $m \times n$ matrix and let $\mathcal{M}$, $\mathcal{N}$, respectively, denote the sets $\{1, \ldots, m\}$, $\{1, \ldots, n\}$. If $\mathcal{M}_1$ is an *ordered* subset of $\mathcal{M}$, i.e. $\mathcal{M}_1 = \{i_1, \ldots, i_r\}$, $i_1 < \cdots < i_r$, and $\mathcal{N}_1 = \{k_1, \ldots, k_s\}$ an ordered subset of $\mathcal{N}$, then $A(\mathcal{M}_1, \mathcal{N}_1)$ denotes the $r \times s$ *submatrix* of A obtained from A by leaving the rows with indices in $\mathcal{M}_1$ and removing all the remaining rows and leaving the columns with indices in $\mathcal{N}_1$ and removing the remaining columns.

Particularly important are submatrices corresponding to consecutive row indices as well as consecutive column indices. Such a submatrix is called a *block* of the original matrix. We then obtain a *partitioning* of the matrix A into blocks by splitting the set of row indices into subsets of the first, say, p_1 indices, then the set of the next p_2 indices, etc., up to the last p_u indices, and similarly splitting the set of column indices into subsets of consecutive

$q_1, \ldots, q_v$ indices. If A_{rs} denotes the block describing the $p_r \times q_s$ submatrix of A obtained by this procedure, A can be written as

$$A = \begin{bmatrix} A_{11} & A_{12} & \ldots & A_{1v} \\ A_{21} & A_{22} & \ldots & A_{2v} \\ \cdot & \cdot & \ldots & \cdot \\ A_{u1} & A_{u2} & \ldots & A_{uv} \end{bmatrix}.$$

If, for instance, we partition the 3×4 matrix $[a_{ik}]$ with $p_1 = 2$, $p_2 = 1$, $q_1 = 1$, $q_2 = 2$, $q_3 = 1$, we obtain the block matrix

$$\begin{bmatrix} A_{11} & A_{12} & A_{13} \\ A_{21} & A_{22} & A_{23} \end{bmatrix},$$

where, say A_{12} denotes the block $\begin{bmatrix} a_{12} & a_{13} \\ a_{22} & a_{23} \end{bmatrix}$.

On the other hand, we can form matrices from blocks. We only have to fulfill the condition that all matrices in each block row must have the same number of rows and all matrices in each block column must have the same number of columns.

The importance of block matrices lies in the fact that we can multiply block matrices in the same way as before:

Let $A = [A_{ik}]$ and $B = [B_{kl}]$ be block matrices, A with m block rows and n block columns, and B with n block rows and p block columns. If (and this is crucial) the first block column of A has the same number of columns as the first block row of B has number of rows, the second block column of A has the same number of columns as the second block row of B has number of rows, etc., then the product $C = AB$ is the matrix $C = [C_{il}]$, where

$$C_{il} = A_{i1}B_{1l} + A_{i2}B_{2l} + \cdots + A_{in}B_{nl}.$$

Now let $A = [a_{ik}]$ be an $m \times n$ matrix. The $n \times m$ matrix $C = [c_{pq}]$ for which $c_{pq} = a_{qp}$, $p = 1, \ldots, n$, $q = 1, \ldots, m$, is called the *transpose matrix* of A. It is denoted by A^T. If A and B are matrices that can be multiplied, then

$$(AB)^T = B^T A^T.$$

Also

$$(A^T)^T = A$$

for every matrix A.

This notation is also advantageous for vectors. We usually denote the column vector u with entries (coordinates) $u_1, \ldots, u_n$ as $[u_1, \ldots, u_n]^T$.

Of crucial importance are square matrices. If of fixed order, say n, and over a fixed field, e.g. R or C, they form a set that is *closed* with respect to addition and multiplication as well as transposition. Here, closed means that the result of the operation again belongs to the set.

A square matrix $A = [a_{ik}]$ of order n is called *diagonal* if $a_{ik} = 0$ whenever $i \neq k$. Such a matrix is usually described by its diagonal entries as $\operatorname{diag}\{a_{11}, \ldots, a_{nn}\}$. The matrix A is called *lower triangular* if $a_{ik} = 0$, whenever $i < k$, and *upper triangular* if $a_{ik} = 0$, whenever $i > k$. We have then:

Observation A.1.1 *The set of diagonal (respectively, lower triangular, respectively, upper triangular) matrices of fixed order over a fixed field R or C is closed with respect to both addition and multiplication.*

A square matrix $A = [a_{ik}]$ is called *tridiagonal* if $a_{ik} = 0$,whenever $|i - k| > 1$; thus only diagonal entries and the entries right above or below the diagonal can be different from zero.

A matrix A (necessarily square!) is called *nonsingular* if there exists a matrix C such that $AC = CA = I$. This matrix C (which can be shown to be unique) is called the *inverse matrix of A* and is denoted by A^{-1}. Clearly

$$(A^{-1})^{-1} = A.$$

Observation A.1.2 *If A, B are nonsingular matrices of the same order, then their product AB is also nonsingular and*

$$(AB)^{-1} = B^{-1}A^{-1}.$$

Observation A.1.3 *If A is nonsingular, then A^T is nonsingular and*

$$(A^T)^{-1} = (A^{-1})^T.$$

Let us recall now the notion of the *determinant* of a square matrix $A = [a_{ik}]$ of order n. We denote it as $\det A$

$$\det A = \sum_{P=(k_1,\ldots,k_n)} \sigma(P) a_{1k_1} a_{2k_2} \cdots a_{nk_n},$$

where the sum is taken over all permutations $P = (k_1, k_2, \ldots, k_n)$ of the indices $1, 2, \ldots, n$, and $\sigma(P)$, the *sign* of the permutation P, is 1 or -1, according to whether the number of pairs (i, j) for which $i < j$ but $k_i > k_j$ is even or odd.

In this connection, let us mention that an $n \times n$ matrix which has in the first row just one nonzero entry 1 in the position $(1, k_1)$, in the second row one nonzero entry 1 in the position $(2, k_2)$ etc., in the last row one nonzero entry 1 in the position (n, k_n) is called a *permutation matrix*. If it is denoted as $\hat{P}$, then

$$\hat{P}\hat{P}^T = I. \tag{A.2}$$

We now list some important properties of the determinants.

Theorem A.1.4 *Let $A = [a_{ik}]$ be a lower triangular, upper triangular, or diagonal matrix of order n. Then*

$$\det A = a_{11}a_{22}\dots a_{nn}.$$

In particular

$$\det I = 1$$

for every identity matrix.

We denote by $|\mathcal{S}|$ the number of elements in a set $\mathcal{S}$. Let A be a square matrix of order n. Denote, as before, $\mathcal{N} = \{1, \dots, n\}$. Whenever $\mathcal{M}_1 \subset \mathcal{N}$, $\mathcal{M}_2 \subset \mathcal{N}$, and $|\mathcal{M}_1| = |\mathcal{M}_2|$, the submatrix $A(\mathcal{M}_1, \mathcal{M}_2)$ is square. We then call $\det A(\mathcal{M}_1, \mathcal{M}_2)$ a *subdeterminant* or *minor* of the matrix A. If $\mathcal{M}_1 = \mathcal{M}_2$, we speak about *principal minors* of A.

Theorem A.1.5 *If P and Q are square matrices of the same order, then*

$$\det PQ = \det P \cdot \det Q.$$

Theorem A.1.6 *A matrix $A = [a_{ik}]$ is nonsingular if and only if it is square and its determinant is different from zero. In addition, the inverse $A^{-1} = [\alpha_{ik}]$ where*

$$\alpha_{ik} = \frac{A_{ki}}{\det A}, \tag{A.3}$$

A_{ki} being the algebraic complement of a_{ki}.

Remark A.1.7 The transpose matrix of the algebraic complements is called the *adjoint matrix* of the matrix A and denoted as adj$A$.

Remark A.1.8 Theorem A.1.5 implies that the product of a finite number of nonsingular matrices of the same order is again nonsingular.

Remark A.1.9 Theorem A.1.6 implies that for checking that the matrix C is the inverse of A, only one of the conditions $AC = I$, $CA = I$ suffices.

Let us return, for a moment, to the block lower triangular matrix as in Observation A.1.1.

Theorem A.1.10 *A block triangular matrix*

$$A = \begin{bmatrix} A_{11} & 0 & 0 & \dots & 0 \\ A_{21} & A_{22} & 0 & \dots & 0 \\ . & . & . & \dots & 0 \\ A_{r1} & A_{r2} & A_{r3} & \dots & A_{rr} \end{bmatrix}$$

with square diagonal blocks is nonsingular if and only if all the diagonal blocks are nonsingular. In such a case the inverse $A^{-1} = [B_{ik}]$ is also lower block *triangular. The diagonal blocks B_{ii} are the inverses of A_{ii} and the subdiagonal blocks B_{ij}, $i > j$, can be obtained recurrently from*

$$B_{ij} = -A_{ii}^{-1} \sum_{k=j}^{i-1} A_{ik} B_{kj}.$$

Remark A.1.11 This theorem applies, of course, also to the simplest case when the blocks A_{ik} are entries of the lower triangular matrix $[a_{ik}]$. An analogous result on inverting upper triangular matrices, or upper block triangular matrices, follows by transposing the matrix and using Observation A.1.3.

A square matrix A of order n is called *strongly nonsingular* if all the principal minors $\det A(\mathcal{N}_k, \mathcal{N}_k)$, $k = 1, \ldots, n$, $\mathcal{N}_k = \{1, \ldots, k\}$ are different from zero.

Theorem A.1.12 *Let A be a square matrix. Then the following are equivalent:*

(i) *A is strongly nonsingular.*
(ii) *A has an LU-decomposition, i.e. there exist a nonsingular lower triangular matrix L and a nonsingular upper triangular matrix U such that $A = LU$.*

The condition (ii) *can be formulated in a stronger form: $A = BDC$, where B is a lower triangular matrix with ones on the diagonal, C is an upper triangular matrix with ones on the diagonal, and D is a nonsingular diagonal matrix. This factorization is uniquely determined. The diagonal entries d_k of D are*

$$d_1 = A(\{1\}, \{1\}), \quad d_k = \frac{\det A(\mathcal{N}_k, \mathcal{N}_k)}{\det A(\mathcal{N}_{k-1}, \mathcal{N}_{k-1})}, \qquad k = 2, \ldots, n.$$

Now let

$$A = \begin{bmatrix} A_{11} & A_{12} \\ A_{21} & A_{22} \end{bmatrix} \tag{A.4}$$

be a block matrix in which A_{11} is nonsingular. We then call the matrix

$$A_{22} - A_{21} A_{11}^{-1} A_{12}$$

the *Schur complement* of the submatrix A_{11} in A and denote it by $[A/A_{11}]$. Here, the matrix A_{22} does not even need to be square.

Theorem A.1.13 *If the matrix*

$$A = \begin{bmatrix} A_{11} & A_{12} \\ A_{21} & A_{22} \end{bmatrix}$$

is square and A_{11} is nonsingular, then the matrix A is nonsingular if and only if the Schur complement $[A/A_{11}]$ is nonsingular. We have then

$$\det A = \det A_{11} \det[A/A_{11}],$$

and if the inverse

$$A^{-1} = \begin{bmatrix} B_{11} & B_{12} \\ B_{21} & B_{22} \end{bmatrix}$$

is written in the same block form, then

$$[A/A_{11}] = B_{22}^{-1}. \tag{A.5}$$

If A is not nonsingular, then the Schur complement $[A/A_{11}]$ is also not nonsingular; if $Az = 0$, then $[A/A_{11}]\hat{z} = 0$, where $\hat{z}$ is the column vector obtained from z by omitting coordinates with indices in A_{11}.

Starting with the inverse matrix, we obtain immediately:

Corollary A.1.14 *The inverse of a nonsingular principal submatrix of a nonsingular matrix is the Schur complement of the inverse with respect to the submatrix with the complementary set of indices. In other words, if both A and A_{11} in*

$$A = \begin{bmatrix} A_{11} & A_{12} \\ A_{21} & A_{22} \end{bmatrix}$$

are nonsingular, then

$$A_{11}^{-1} = [A^{-1}/(A^{-1})_{22}]. \tag{A.6}$$

Remark A.1.15 The principal submatrices of a square matrix enjoy the property that if A_1 is a principal submatrix of A_2 and A_2 is a principal submatrix of A_3, then A_1 is a principal submatrix of A_3 as well. This property is essentially reflected, in the case of a nonsingular matrix, thanks to Corollary A.1.14, by the Schur complements: if

$$A = \begin{bmatrix} A_{11} & A_{12} & A_{13} \\ A_{21} & A_{22} & A_{23} \\ A_{31} & A_{32} & A_{33} \end{bmatrix}$$

and

$$\begin{bmatrix} A_{11} & A_{12} \\ A_{21} & A_{22} \end{bmatrix}$$

is denoted as $\tilde{A}$, then

$$[A/\tilde{A}] = [[A/A_{11}]/[\tilde{A}/A_{11}]]. \tag{A.7}$$

Let us also mention the *Sylvester identity* in the simplest case which shows how the principal minors of two inverse matrices are related (cf. [31], p. 21):

Theorem A.1.16 *Let again*

$$A = \begin{bmatrix} A_{11} & A_{12} \\ A_{21} & A_{22} \end{bmatrix}$$

be a nonsingular matrix with the inverse

$$A^{-1} = \begin{bmatrix} B_{11} & B_{12} \\ B_{21} & B_{22} \end{bmatrix}.$$

Then

$$\det B_{11} = \frac{\det A_{22}}{\det A}. \tag{A.8}$$

If V_1 is a nonempty subset in a vector space V, which is closed with respect to the operations of addition and scalar multiplication in V, then we say that V_1 is a *linear subspace* of V. It is clear that the intersection of linear subspaces of V is again a linear subspace of V. In this sense, the set (0) is in fact a linear subspace contained in all linear subspaces of V.

If S is some set of vectors of a finite-dimensional vector space V, then the linear subspace of V of smallest dimension that contains the set S is called the *linear hull* of S and its dimension (necessarily finite) is called the *rank* of S.

We are now able to present, without proof, an important statement about the *rank* of a matrix.

Theorem A.1.17 *Let A be an $m \times n$ matrix. Then the rank of the system of the columns (as vectors) of A is the same as the rank of the system of the rows (as vectors) of A. This common number $r(A)$, called the rank of the matrix A, is equal to the maximum order of all nonsingular submatrices of A. (If A is the zero matrix, thus containing no nonsingular submatrix, then $r(A) = 0$.)*

Theorem A.1.18 *A square matrix A is singular if and only if there exists a nonzero vector x for which $Ax = 0$.*

The rank function enjoys important properties. We list some:

Theorem A.1.19 *We have:*

(i) *For any matrix A,*

$$r(A^T) = r(A).$$

(ii) *If the matrices A and B are of the same type, then*

$$r(A + B) \leq r(A) + r(B).$$

(iii) *If the matrices A and B can be multiplied, then*

$$r(AB) \leq \min(r(A), r(B)).$$

(iv) *If A (respectively, B) is nonsingular, then $r(AB) = r(B)$ (respectively, $r(AB) = r(A)$).*

(v) *If a matrix A has rank one, then there exist nonzero column vectors x and y such that $A = xy^T$.*

Theorem A.1.13 can now be completed as follows:

Theorem A.1.20 *In the same notation as in* (A.4)

$$r(A) = r(A_{11}) + r([A/A_{11}]).$$

For square matrices, the following important notions have to be mentioned.

Let A be a square matrix of order n. A nonzero column vector x is called an *eigenvector* of A if $Ax = \lambda x$ for some number (scalar) λ. This number λ is called an *eigenvalue* of A corresponding to the eigenvector x.

Theorem A.1.21 *A necessary and sufficient condition that a number λ is an eigenvalue of a matrix A is that the matrix $A - \lambda I$ is singular, i.e. that*

$$\det(A - \lambda I) = 0.$$

This formula is equivalent to

$$(-\lambda)^n + c_1(-\lambda)^{n-1} + \cdots + c_{n-1}(-\lambda) + c_n = 0, \tag{A.9}$$

where c_k is the sum of all principal minors of A of order k

$$c_k = \sum_{\mathcal{M} \subset \mathcal{N},\ |\mathcal{M}|=k} \det A(\mathcal{M}, \mathcal{M}),\ \mathcal{N} = \{1, \ldots, n\}.$$

The polynomial on the left-hand side of (A.9) is called the *characteristic polynomial* of the matrix A. It has degree n.

We have thus:

Theorem A.1.22 *A square complex matrix $A = [a_{ik}]$ of order n has n eigenvalues (some may coincide). These are all the roots of the characteristic polynomial of A. If we denote them as $\lambda_1, \ldots, \lambda_n$, then*

$$\sum_{i=1}^{n} \lambda_i = \sum_{i=1}^{n} a_{ii}, \tag{A.10}$$

$$\lambda_1 \lambda_2 \cdot \ldots \cdot \lambda_n = \det A.$$

The number $\sum_{i=1}^{n} a_{ii}$ is called the *trace* of the matrix A. We denote it by $\operatorname{tr} A$. By (A.10), $\operatorname{tr} A$ is the sum of all the eigenvalues of A.

Remark A.1.23 A real square matrix need not have real eigenvalues, but as its characteristic polynomial has real coefficients, the nonreal eigenvalues occur in complex conjugate pairs.

Theorem A.1.24 *A real or complex square matrix is nonsingular if and only if all its eigenvalues are different from zero. In such a case, the inverse has eigenvalues reciprocal to the eigenvalues of the matrix.*

We can also use the more general Gaussian block elimination method.

Theorem A.1.25 *Let the system $Ax = b$ be in the block form*

$$\begin{aligned} A_{11}x_1 + A_{12}x_2 &= b_1, \\ A_{21}x_1 + A_{22}x_2 &= b_2, \end{aligned}$$

where x_1, x_2 are vectors.

If A_{11} is nonsingular, then this system is equivalent to the system

$$\begin{aligned} A_{11}x_1 + A_{12}x_2 &= b_1, \\ (A_{22} - A_{21}A_{11}^{-1}A_{12})x_2 &= b_2 - A_{21}A_{11}^{-1}b_1. \end{aligned}$$

Remark A.1.26 In this theorem, the role of the Schur complement $[A/A_{11}] = A_{22} - A_{21}A_{11}^{-1}A_{12}$ for elimination is recognized.

Now, we pay attention to a specialized real vector space called *Euclidean vector space* in which magnitude (length) of a vector is defined by the so-called *inner product* of two vectors.

For the sake of completeness, recall that a real finite-dimensional vector space E is called a *Euclidean vector space* if a function $\langle x, y\rangle : E \times E \to R$ is given that satisfies:

E 1. $\langle x, y\rangle = \langle y, x\rangle$ for all $x \in E$, $y \in E$;
E 2. $\langle x_1 + x_2, y\rangle = \langle x_1, y\rangle + \langle x_2, y\rangle$ for all $x_1 \in E$, $x_2 \in E$, and $y \in E$;
E 3. $\langle \alpha x, y\rangle = \alpha\langle x, y\rangle$ for all $x \in E$, $y \in E$, and all real α;
E 4. $\langle x, x\rangle \geq 0$ for all $x \in E$, with equality if and only if $x = 0$.

The property E 4 enables us to define the *length* $\|x\|$ of the vector x as $\sqrt{\langle x, x\rangle}$. A vector is called a *unit vector* if its length is one. Vectors x and y are *orthogonal* if $\langle x, y\rangle = 0$. A system $u_1, \dots, u_m$ of vectors in E is called *orthonormal* if $\langle u_i, u_j\rangle = \delta_{ij}$, the Kronecker delta.

It is easily proved that every orthonormal system of vectors is linearly independent. If the number of vectors in such a system is equal to the dimension of E, it is called an *orthonormal basis* of E.

The real vector space R^n of column vectors will become a Euclidean space if the inner product of the vectors $x = [x_1, \dots, x_n]^T$ and $y = [y_1, \dots, y_n]^T$ is defined as

$$\langle x, y\rangle = x_1y_1 + \cdots + x_ny_n;$$

in other words, in matrix notation

$$\langle x, y\rangle = x^Ty \ \ (= y^Tx). \tag{A.11}$$

Notice the following:

Theorem A.1.27 *If a vector is orthogonal to all the vectors of a basis, then it is the zero vector.*

We call now a matrix $A = [a_{ik}]$ in $R^{n\times n}$ *symmetric* if $a_{ik} = a_{ki}$ for all i, k, or equivalently, if $A = A^T$. We call the matrix A *orthogonal* if $AA^T = I$. Thus:

Theorem A.1.28 *The sum of two symmetric matrices in $R^{n\times n}$ is symmetric; the product of two orthogonal matrices in $R^{n\times n}$ is orthogonal. The identity matrix is orthogonal and the transpose (which is equal to the inverse) of an orthogonal matrix is orthogonal as well.*

The following theorem on orthogonal matrices holds (see [28]):

Theorem A.1.29 *Let Q be an $n \times n$ real matrix. Then the following are equivalent:*

(i) *Q is orthogonal.*
(ii) *For all $x \in R^n$*
$$\|Qx\| = \|x\|.$$
(iii) *For all $x \in R^n$, $y \in R^n$*
$$\langle Qx, Qy\rangle = \langle x, y\rangle.$$
(iv) *Whenever $u_1, \ldots, u_n$ is an orthonormal basis, then $Qu_1, \ldots, Qu_n$ is an orthonormal basis as well.*
(v) *There exists an orthonormal basis $v_1, \ldots, v_n$ such that $Qv_1, \ldots, Qv_n$ is again an orthonormal basis.*

The basic theorem on symmetric matrices can be formulated as follows.

Theorem A.1.30 *Let A be a real symmetric matrix. Then there exist an orthogonal matrix Q and a real diagonal matrix D such that $A = QDQ^T$. The diagonal entries of D are the eigenvalues of A, and the columns of Q are eigenvectors of A; the kth column corresponds to the kth diagonal entry of D.*

Theorem A.1.31 *All the eigenvalues of a real symmetric matrix are real. For every real symmetric matrix, there exists an orthonormal basis of R consisting of its eigenvectors.*

If a real symmetric matrix A has p positive and q negative eigenvalues, then the difference $p - q$ is called the *signature* of the matrix A. By the celebrated Inertia Theorem the following holds (cf. [28], Theorem 2.6.1):

Theorem A.1.32 *The signatures of a real symmetric matrix A and CAC^T are equal whenever C is a real nonsingular matrix of the same type as A.*

We also should mention the theorem on the interlacing of eigenvalues of submatrices (cf. [31], p.185).

Theorem A.1.33 *Let $A \in R^{n\times n}$ be symmetric, $y \in R^n$ and α a real number. Let $\tilde{A}$ be the matrix*

$$\tilde{A} = \begin{bmatrix} A & y \\ y^T & \alpha \end{bmatrix}.$$

If $\lambda_1 \le \lambda_2 \le \cdots \le \lambda_n$ (respectively, $\tilde{\lambda}_1 \le \tilde{\lambda}_2 \le \cdots \le \tilde{\lambda}_{n+1}$) are the eigenvalues of A (respectively, $\tilde{A}$), then

$$\tilde{\lambda}_1 \le \lambda_1 \le \tilde{\lambda}_2 \le \lambda_2 \le \cdots \le \tilde{\lambda}_n \le \lambda_n \le \tilde{\lambda}_{n+1}.$$

An important subclass of the class of real symmetric matrices is that of positive definite (respectively, positive semidefinite) matrices.

A real symmetric matrix A of order n is called *positive definite* (respectively, *positive semidefinite*) if for every nonzero vector $x \in R^n$, the product $x^T Ax$ is positive (respectively, nonnegative).

In the following theorem we collect the basic *characteristic* properties of positive definite matrices. For the proof, see [28].

Theorem A.1.34 *Let $A = [a_{ik}]$ be a real symmetric matrix of order n. Then the following are equivalent:*

(i) *A is positive definite.*
(ii) *All principal minors of A are positive.*
(iii) (Sylvester criterion) $\det A(\mathcal{N}_k, \mathcal{N}_k) > 0$ *for $k = 1, \ldots, n$, where $\mathcal{N}_k = \{1, \ldots, k\}$. In other words*

$$a_{11} > 0, \ \det \begin{bmatrix} a_{11} & a_{12} \\ a_{21} & a_{22} \end{bmatrix} > 0, \ \ldots, \ \det A > 0.$$

(iv) *There exists a nonsingular lower triangular matrix B such that $A = BB^T$.*
(v) *There exists a nonsingular matrix C such that $A = CC^T$.*
(vi) *The sum of all the principal minors of order k is positive for $k = 1, \ldots, n$.*
(vii) *All the eigenvalues of A are positive.*
(viii) *There exists an orthogonal matrix Q and a diagonal matrix D with positive diagonal entries such that $A = QDQ^T$.*

Corollary A.1.35 *If A is positive definite, then A^{-1} exists and is positive definite as well.*

Remark A.1.36 Observe also that the identity matrix is positive definite.

For positive semidefinite matrices, we have:

Theorem A.1.37 *Let $A = [a_{ik}]$ be a real symmetric matrix of order n. Then the following are equivalent:*

(i) *A is positive semidefinite.*
(ii) *The matrix $A + \varepsilon I$ is positive definite for all $\varepsilon > 0$.*

(iii) *All the principal minors of* A *are nonnegative.*
(iv) *There exists a square matrix* C *such that* $A = CC^T$.
(v) *The sum of all principal minors of order* k *is nonnegative for* $k = 1, \ldots, n$.
(vi) *All the eigenvalues of* A *are nonnegative.*
(vii) *There exists an orthogonal matrix* Q *and a diagonal matrix* D *with nonnegative diagonal entries such that* $A = QDQ^T$.

Corollary A.1.38 *A positive semidefinite matrix is positive definite if and only if it is nonsingular.*

Corollary A.1.39 *If* A *is positive definite and* α *a positive number, then* αA *is positive definite as well. If* A *and* B *are positive definite of the same order, then* $A + B$ *is positive definite; this is so even if one of the matrices* A, B *is positive semidefinite.*

The expression $x^T Ax$ – in the case that A is symmetric – is called the *quadratic form* corresponding to the matrix A. If A is positive definite (respectively, positive semidefinite), this quadratic form is called positive definite (respectively, positive semidefinite). Observe also the following property:

Theorem A.1.40 *If* A *is a positive semidefinite matrix and* y *a real vector for which* $y^T Ay = 0$, *then* $Ay = 0$.

Remark A.1.41 Theorem A.1.37 can be also formulated using the rank r of A in the separate items. So in (iv), the matrix C can be taken as an $n \times r$ matrix, in (vii) the diagonal matrix D can be specified as having r positive and $n - r$ zero diagonal entries, etc.

We should not forget to mention important inequalities for positive definite matrices. In the first, we use the symbol N for the index set $\{1, 2, \ldots, n\}$ and $A(M)$ for the principal submatrix of A with index set M. For M void, we should put $\det A(M) = 1$.

Theorem A.1.42 (Generalized Hadamard inequality (cf. [31]), p. 478) *Let* A *be a positive definite* $n \times n$ *matrix. Then for any* $M \subset N$

$$\det A \leq \det A(M) \det A(N \backslash M). \tag{A.12}$$

Remark A.1.43 Equality in (A.12) is attained if and only if all entries of A with one index in M and the second in $N \backslash M$ are equal to zero. A further generalization of (A.12) is the Hadamard–Fischer inequality (cf. [31], p. 485)

$$\det A(N_1 \cup N_2) \det A(N_1 \cap N_2) \leq \det A(N_1) \det A(N_2) \tag{A.13}$$

for any subsets $N_1 \subset N$, $N_2 \subset N$.

Concluding this section, let us notice a close relationship of the class of positive semidefinite matrices with Euclidean geometry. If $v_1, v_2, \ldots, v_m$ is a system of vectors in a Euclidean vector space, then the matrix of the inner products

$$G(v_1, v_2, \ldots, v_m) = \begin{bmatrix} \langle v_1, v_1\rangle & \langle v_1, v_2\rangle & \cdots & \langle v_1, v_m\rangle \\ \langle v_2, v_1\rangle & \langle v_2, v_2\rangle & \cdots & \langle v_2, v_m\rangle \\ \cdot & \cdot & \cdot & \cdot \\ \langle v_m, v_1\rangle & \langle v_m, v_2\rangle & \cdots & \langle v_m, v_m\rangle \end{bmatrix}, \tag{A.14}$$

the so-called *Gram matrix* of the system, enjoys the following properties. (Because of its importance for our approach, we supply proofs of the next four theorems.)

Theorem A.1.44 *Let $a_1, a_2, \ldots, a_s$ be vectors in E_n. Then the Gram matrix $G(a_1, a_2, \ldots, a_s)$ is positive semidefinite. It is nonsingular if and only if $a_1, \ldots, a_s$ are linearly independent.*

If $G(a_1, a_2, \ldots, a_s)$ is singular, then every linear dependence relation

$$\sum_{i=1}^{s} \alpha_i a_i = 0 \tag{A.15}$$

implies the same relation among the columns of the matrix $G(a_1, \ldots, a_s)$, i.e.

$$G(a_1, \ldots, a_s)[\alpha] = 0 \text{ for } [\alpha] = [\alpha_1, \ldots, \alpha_s]^T, \tag{A.16}$$

and conversely, every linear dependence relation (A.16) *among the columns of $G(a_1, \ldots, a_s)$ implies the same relation* (A.15) *among the vectors $a_1, \ldots, a_s$.*

Proof. Positive semidefiniteness of $G(a_1, \ldots, a_s)$ follows from the fact that for $x = [x_1, \ldots, x_s]^T$, the corresponding quadratic form $x^T G(a_1, \ldots, a_s)x$ is equal to the inner product $\langle \sum_{i=1}^{s} x_i a_i, \sum_{i=1}^{s} x_i a_i \rangle$, which is nonnegative. In addition, if the vectors $a_1, \ldots, a_s$ are linearly independent, this inner product is positive unless x is zero. Now let (A.15) be fulfilled. Then

$$G(a_1, \ldots, a_s)[\alpha] = \begin{bmatrix} \langle \sum_{i=1}^{s} \alpha_i a_i, a_1 \rangle \\ \vdots \\ \langle \sum_{i=1}^{s} \alpha_i a_i, a_s \rangle \end{bmatrix},$$

which is the zero vector.

Conversely, if (A.16) is fulfilled, then

$$[\alpha]^T G(a_1, \ldots, a_s)[\alpha] = \left\langle \sum_{i=1}^{s} \alpha_i a_i, \sum_{i=1}^{s} \alpha_i a_i \right\rangle$$

is zero. Thus we obtain (A.15). □

Theorem A.1.45 *Every positive semidefinite matrix is the Gram matrix of some system of vectors in any Euclidean space, the dimension of which is greater than or equal to the rank of the matrix.*

Proof. Let A be such a matrix, and let r be its rank. By Remark A.1.41, in (iv) of Theorem A.1.37, A can be written as CC^T, where C is a (real) $n \times r$-matrix. If the rows of C are considered as coordinates of vectors in an r-dimensional (or, higher-dimensional) Euclidean space, then A is the Gram matrix of these points. □

This theorem can be used for obtaining matrix inequalities (cf. [9]).

Theorem A.1.46 *Let $A = [a_{ik}]$ be a positive semidefinite $n \times n$ matrix with row sums zero. Then the square roots of the diagonal entries of A satisfy the polygonal inequality*

$$2 \max_i \sqrt{a_{ii}} \leq \sum_i \sqrt{a_{ii}}. \tag{A.17}$$

Proof. By Theorem A.1.45, A is the Gram matrix of a system of vectors $u_1, \ldots, u_n$ whose sum is the zero vector, thus forming a closed polygon. Therefore, the length $|u_i|$, which is $\sqrt{a_{ii}}$, is less than or equal to the sum of the lengths of the remaining vectors. Consequently, (A.17) follows. □

Theorem A.1.47 *Let $a_1, \ldots, a_n$ be some basis of an n-dimensional Euclidean space E_n. Then there exists a unique ordered system of vectors $b_1, \ldots, b_n$ in E_n for which*

$$\langle a_i, b_k \rangle = \delta_{ik}, \quad i, k = 1, \ldots, n. \tag{A.18}$$

The Gram matrices $G(a_1, \ldots, a_n)$ and $G(b_1, \ldots, b_n)$ are inverse to each other.

Proof. Uniqueness: Suppose that $b'_1, \ldots, b'_n$ also satisfy (A.18). Then for $k = 1, \ldots, n$, the vector $b_k - b'_k$ is orthogonal to all the vectors a_i, hence to all the vectors in $\mathcal{E}_n$. By Theorem A.1.27, $b'_k = b_k$, $k = 1, \ldots, n$.

To prove the existence of $b_1, \ldots, b_n$, denote by $G(a)$ the Gram matrix $G(a_1, \ldots, a_n)$. Observe that the fact that a vector b satisfies the linear dependence relation

$$b = x_1 a_1 + x_2 a_2 + \cdots + x_n a_n \tag{A.19}$$

is equivalent to

$$\begin{bmatrix} \langle a_1, b \rangle \\ \langle a_2, b \rangle \\ \vdots \\ \langle a_n, b \rangle \end{bmatrix} = G(a) \begin{bmatrix} x_1 \\ x_2 \\ \vdots \\ x_n \end{bmatrix}.$$

Thus the vectors b_i, $i = 1, 2. \ldots, n$, defined by (A.19) for $x = G(a)^{-1}e_i$, where $e_i = [\delta_{i1}, \delta_{i2}, \ldots, \delta_{in}]^T$, satisfy (A.18). For fixed i and x for $b = b_i$, we obtain from (A.19), using inner multiplication by b_j and (A.18), that $x_j = \langle b_i, b_j \rangle$. Therefore, using again inner multiplication of (A.19) by a_j, we obtain $I = G(a)G(b_1, b_2, \ldots, b_n)$. □

The two bases $a_1, \ldots, a_n$ and $b_1, \ldots, b_n$ satisfying (A.18) are called *biorthogonal bases.*

As we know, the cosine of the angle φ of two non-zero vectors u_1, u_2 satisfies

$$\cos\varphi = \frac{\langle u_1, u_2 \rangle}{\sqrt{\langle u_1, u_1 \rangle}\sqrt{\langle u_2, u_2 \rangle}}.$$

Therefore

$$\sin^2\varphi = \frac{1}{\langle u_1, u_1 \rangle \langle u_2, u_2 \rangle} \det G(u_1, u_2),$$

where $G(u_1, u_2)$ is the Gram matrix of u_1 and u_2.

Formally, this notion can be generalized for the case of more than two nonzero vectors. Thus, the number denoted as

$$\sin(u_1, \ldots, u_n) = \sqrt{\frac{\det G(u_1, \ldots, u_n)}{\langle u_1, u_1 \rangle \cdots \langle u_n, u_n \rangle}},$$

which is always less than or equal to 1– is called the *sine*, sometimes *spatial sine*, of the vectors u_1,..., u_n.

We add a few words on measuring. The determinant of the Gram matrix $G(u_1, \ldots, u_n)$ is considered as the square of the n-dimensional volume of the parallelepiped spanned by the vectors $u_1, \ldots, u_n$. On the other hand, the volume of the unit cube is an $n!$-multiple of the volume of the special Schlaefli right n-simplex with unit legs since the cube can be put together from that number of such congruent n-simplexes. By affine transformations, it follows that the volume of the mentioned parallelepiped is the $n!$-multiple of the volume V_Σ of the n-simplex Σ with vertices $A_1, A_2, \ldots, A_{n+1}$, where A_{n+1} is a chosen point in $\mathcal{E}_n$ and the A_is are chosen by $A_i = A_{n+1} + u_i$, $i = 1, \ldots, n$.

We now intend to relate the determinant of the extended Menger matrix of Σ with the determinant of the Gram matrix $G(u_1, \ldots, u_n)$. Thus, as in (1.19), let m_{ik} denote the square of the length of the edge A_iA_k for $i \neq k$. We have then $\langle u_i, u_i \rangle = m_{i,n+1}$, and, since $\langle u_i - u_k, u_i - u_k \rangle = m_{ik}$, we obtain $\langle u_i, u_k \rangle = \frac{1}{2}(m_{i,n+1} + m_{k,n+1} - m_{ik})$. Multiplying each row of $G(u_1, \ldots, u_n)$ by 2, we obtain for the determinant

$$\det G(u_1, \ldots, u_n) = \frac{1}{2^n} \det Z,$$

where $Z =$

$$\begin{bmatrix} 2m_{1,n+1} & m_{1,n+1}+m_{2,n+1}-m_{12} & \cdots & m_{1,n+1}+m_{n,n+1}-m_{1n} \\ m_{2,n+1}+m_{1,n+1}-m_{12} & 2m_{2,n+1} & \cdots & m_{2,n+1}+m_{n,n+1}-m_{2n} \\ \vdots & \vdots & \ddots & \vdots \\ m_{n,n+1}+m_{1,n+1}-m_{1n} & m_{n,n+1}+m_{2,n+1}-m_{2n} & \cdots & 2m_{n,n+1} \end{bmatrix}.$$

Bordering now Z by a column with entries $-m_{i,n+1}$ and another column of all ones, and by two rows, the first with $n+1$ zeros and a 1, the second with n zeros, then 1 and 0, the determinant of Z just changes the sign. Add now the $(n+1)$th column to each of the first n columns and subtract the $m_{k,n+1}$-multiple of the last column from the kth column for $k = 1, \ldots, n$. We obtain the extended Menger matrix of Σ in which each entry m_{ik} is multiplied by -1. Its determinant is thus $(-1)^n \det M_0$ in the notation of (1.26).

Altogether, we arrive at

$$\begin{aligned} V_\Sigma^2 &= \left(\frac{1}{n!}\right)^2 \det G(u_1, \ldots, u_n) \\ &= \frac{(-1)^{n+1}}{(n!)^2 2^n} \det M_0, \end{aligned}$$

as was claimed in (1.28).

A.2 Graphs and matrices

A (finite) *directed graph* $G = (V, E)$ consists of the *set of nodes* V and the *set of arcs* E, a subset of the cartesian product $V \times V$. This means that every edge is an ordered pair of nodes and can thus be depicted in the plane by an arc with an arrow if the nodes are depicted as points.

For our purpose, it will be convenient to choose V as the set $\{1, 2, \ldots, n\}$ (or, to *order* the set V). If now E is the set of arcs of G, define an $n \times n$ matrix $A(G)$ as follows: if there is an arc "starting" in i and "ending" in k, the entry in the position (i, k) will be one, if there is no arc starting in i and ending in k, the entry in the position (i, k) will be zero.

We have thus assigned to a finite directed graph (usually called a *digraph*) a (0, 1) matrix $A(G)$. Conversely, let $C = [c_{ik}]$ be an $n \times n$ (say, real) matrix. We can assign to C a digraph $G(C) = (V, E)$ as follows: V is the set $\{1, \ldots, n\}$, and E the set of all pairs (i, k) for which c_{ik} is different from zero.

The graph theory terminology speaks about a *walk* in G from node i to the node k if there are nodes $j_1, \ldots, j_s$ such that all the arcs $(i, j_1), (j_1, j_2), \ldots, (j_s, k)$ are in E; $s+1$ is then the *length* of this walk. The nodes in the walk need not be distinct. If they are, the walk is a *path*. If i coincides with k, we speak about a *cycle*; its length is then again $s+1$. If all the remaining nodes are distinct, the cycle is *simple*. The arcs (k, k) themselves are called *loops*. The digraph is *strongly connected* if there is at least one path from each node to any other node. There is an equivalent property for matrices.

Let P be a *permutation matrix*. By (A.2), we have $PP^T = I$. If C is a square matrix and P a permutation matrix of the same order, then PCP^T is obtained from C by a *simultaneous permutation* of rows *and* columns; the diagonal entries remain diagonal. Observe that the digraph $G(PCP^T)$ differs from the digraph $G(C)$ only by different numbering of the nodes.

We say that a square matrix C is *reducible* if it has the block form

$$C = \begin{bmatrix} C_{11} & C_{12} \\ 0 & C_{22} \end{bmatrix},$$

where both matrices C_{11}, C_{22} are square of order at least one, or if it can be brought to such form by a simultaneous permutation of rows and columns.

A matrix is called *irreducible* if it is square and not reducible. (Observe that a 1×1 matrix is always irreducible, even if the entry is zero.)

This relatively complicated notion is important for (in particular, nonnegative) matrices and their applications, e.g. in probability theory. However, it has a very simple equivalent in the graph-theoretical setting.

Theorem A.2.1 *A matrix C is irreducible if and only if the digraph $G(C)$ is strongly connected.*

A more detailed view is given in the following theorem.

Theorem A.2.2 *Every square real matrix can be brought by a simultaneous permutation of rows and columns to the form*

$$\begin{bmatrix} C_{11} & C_{12} & C_{13} & \dots & C_{1r} \\ 0 & C_{22} & C_{23} & \dots & C_{2r} \\ 0 & 0 & C_{33} & \dots & C_{3r} \\ . & . & . & \dots & . \\ 0 & 0 & 0 & \dots & C_{rr} \end{bmatrix},$$

in which the diagonal blocks are irreducible (thus square) matrices.

This theorem has a counterpart in graph theory. Every finite digraph has the following structure. It consists of so-called *strong components* that are the maximal strongly connected subdigraphs; these can then be numbered in such a way that there is no arc from a node with a larger number of the strong component into a node belonging to the strong component with a smaller number.

A digraph is *symmetric* if to every arc (i, j) in E the arc (j, i) is also in E. Such a symmetric digraph can be simply treated as an undirected graph. In graph theory, a *finite undirected graph* (or briefly *graph*) $G = (V, H)$ is introduced as an ordered pair of two finite sets (V, H), where V is the set of nodes and H is the set of some *unordered* pairs of the elements of V, which will here be called *edges*. A finite undirected graph can also be represented by means of a plane diagram in such a way that the nodes of the graph are represented by points in the plane and edges of the graph by segments (or, arcs) joining the corresponding two (possibly also identical) points in the plane. In contrast to the representation of digraphs, the edges are not equipped with arrows.

It is usually required that an undirected graph contains neither *loops* (i.e., the edges (u, u) where $u \in V$), nor more edges joining the same pair of nodes (the so-called *multiple edges*).

If (u, v) is an edge of a graph, we say that this edge is *incident* with the nodes u and v or that the nodes u and v are incident with this edge. In a graph containing no loops, a node is said to have *degree k* if it is incident exactly with k edges. The nodes of degree 0 are called *isolated*; the nodes of degree 1 are called *end-nodes*. An edge incident with an end-node is called a *pending edge*.

We have introduced the concepts of a (directed) walk and a (directed) path in digraphs. Analogous concepts in undirected graphs are a walk and a path. A *walk* in a graph G is a sequence of nodes (not necessarily distinct), say $(u_1, u_2, \ldots, u_s)$ such that every two consecutive nodes u_k and u_{k+1} $(k = 1, \ldots, s-1)$ are joined by an edge in G. A *path* in a graph G is then such a walk in which all the nodes are distinct. A *polygon*, or a *circuit* in G, is a walk whose first and last nodes are identical and, if the last node is removed, all the remaining nodes are distinct. At the same time, this first (and also last) node of the walk representing a circuit is not considered distinguished in the circuit.

We also speak about a *subgraph* of a given graph and about a *union* of graphs. A *connected graph* is defined as a graph in which there exists a path between any two distinct nodes.

If the graph G is not connected, we introduce the notion of a *component* of G as such a subgraph of G which is connected but is not contained in any other connected subgraph of G.

With connected graphs, it is important to study the question of how connectivity changes when some edge is removed (the set of nodes remaining the same), or when some node as well as all the edges incident with it are removed. An edge of a graph is called a *bridge* if it is not a pending edge and if the graph has more components after removing this edge. A node of a connected graph such that the graph has again more components after removing this node (together with all incident edges) is called a *cut-node*. More generally, we call a subset of nodes whose removal results in a disconnected graph a *cut-set* of the graph, for short a *cut*.

The following theorems are useful for the study of cut-nodes and connectivity in general.

Theorem A.2.3 *If a longest path in the graph G joins the nodes u and v, then neither u nor v are cut-nodes.*

Theorem A.2.4 *A connected graph with n nodes, without loops and multiple edges, has at least $n-1$ edges. If it has more than $n-1$ edges, it contains a circuit as a subgraph.*

We now present a theorem on an important type of connected graph.

Theorem A.2.5 *Let G be a connected graph, without loops and multiple edges, with n nodes. Then the following conditions are equivalent:*

(i) *The graph G has exactly $n-1$ edges.*
(ii) *Each edge of G is either a pending edge, or a bridge.*
(iii) *There exists one and only one path between any two distinct nodes of G.*
(iv) *The graph G contains no circuit as a subgraph, but adding any new edge (and no new node) to G, we always obtain a circuit.*
(v) *The graph G contains no circuit.*

A connected graph satisfying one (and then all) of the conditions (i) to (v) of Theorem A.2.5 is called a *tree.*

Every path is a tree; another example of a tree is a *star*, i.e. a graph with n nodes, $n-1$ of which are end-nodes, and the last node is joined with all these end-nodes.

A graph, every component of which is a tree, is called a *forest.*

A subgraph of a connected graph G which has the same vertices as G and which is a tree is called a *spanning tree* of G.

Theorem A.2.6 *There always exists a spanning tree of a connected graph. Moreover, choosing an arbitrary subgraph S of a connected graph G that contains no polygon, we can find a spanning tree of G that contains S as a subgraph.*

Some special graphs should be mentioned. In addition to the path and the circuit, a *wheel* is a graph consisting of a circuit and an additional node which is joined by an edge to each node of the circuit.

An important notion is the *edge connectivity* of a graph. It is the smallest number of edges whose removal causes the graph to be disconnected, or to have only a node left. Clearly, the edge-connectivity of a not-connected graph is zero, the edge connectivity of a tree is one, of a circuit two, and of a wheel (with at least four nodes) three.

Weighted graphs (more precisely, edge-weighted graphs) are graphs in which to every edge (in the case of directed graphs, to every arc) a nonnegative number, called *weight*, is assigned. In such a case, the degree of a node in an undirected graph is the sum of the weights of all the edges incident with that node. Usually, edges with zero weight are considered as "not edges," sometimes missing edges are considered as edges with zero weight. A path is then only such a path in which all the edges have a positive weight. The *length* of a path is then the sum of the weights on all the edges. The distance between two nodes in an undirected graph is the length of a shortest path between those nodes.

Signed graphs are undirected graphs in which every edge is considered either as positive, or as negative. A *signed graph of an* $n \times n$ *real matrix* $A = [a_{ik}]$ is the signed graph on n nodes $1, 2, \ldots, n$ which has positive edges (i, k) if $a_{ik} > 0$ and negative edges (i, k) if $a_{ik} < 0$. The entries a_{ii} are usually not involved. We can then speak about the *positive part* of the graph and *negative part* of the graph. Both have the same set of nodes and the first has just the positive and the second just the negative edges.

A.3 Nonnegative matrices, M- and P-matrices

Positivity, or, more generally, nonnegativity, plays an important role in most parts of this book. In the present section, we always assume that the vectors and matrices are *real*.

We denote by the symbols $>$, $\geq$ or $<$, $\leq$ *componentwise comparison* of the vectors or matrices. For instance, for a matrix A, $A > 0$ means that *all* the entries of A are *positive*; the matrix is called *positive*. $A \geq 0$ means nonnegativity of all the entries and the matrix is called *nonnegative*.

Evidently, the sum of two or more nonnegative matrices of the same type is again nonnegative, and also the product of nonnegative matrices, if they can be multiplied, is nonnegative. Sometimes it is necessary to know whether the result is already positive. Usually, the combinatorial structure of zero and nonzero entries and not the values themselves decide. In such a case, it is useful to apply graph theory terminology. We restrict ourselves to the case of square matrices.

Let us now formulate the *Perron–Frobenius theorem* [28] on nonnegative matrices.

Theorem A.3.1 *Let A be a nonnegative irreducible square matrix of order n, $n > 1$. Then there exists a positive eigenvalue p of A which is simple and such that no other eigenvalue has a greater modulus. There is a positive eigenvector associated with the eigenvalue p and no nonnegative eigenvector is associated with any other eigenvalue of A.*

There is another important class of matrices that is closely related to the previous class of nonnegative matrices.

A square matrix A is called an *M-matrix* if it has the form $kI - C$, where C is a nonnegative matrix and $k > \varrho(C)$, the *spectral radius* of C, i.e. the maximum modulus of all the eigenvalues of C.

Observe that every M-matrix has all the off-diagonal entries non-positive. It is usual to denote the set of such matrices by Z. To characterize matrices from Z to obtain M-matrices, there exist surprisingly many possibilities. We list some:

Theorem A.3.2 *Let A be a matrix in Z of order n. Then the following are equivalent.*

(i) *A is an M-matrix.*
(ii) *There exists a vector $x \geq 0$ such that $Ax > 0$.*
(iii) *All the principal minors of A are positive.*
(iv) *The sum of all the principal minors of order k is positive for $k = 1, \ldots, n$.*
(v) $\det A(\mathcal{N}_k, \mathcal{N}_k) > 0$ *for $k = 1, \ldots, n$, where $\mathcal{N}_k = \{1, \ldots, k\}$.*
(vi) *Every real eigenvalue of A is positive.*
(vii) *The real part of every eigenvalue of A is positive.*
(viii) *A is nonsingular and A^{-1} is nonnegative.*

The proof and other characteristic properties can be found in [28].

Corollary A.3.3 *Let A be an M-matrix. Then every principal submatrix of A as well as the Schur complement of every principal submatrix of A is again an M-matrix.*

Remark A.3.4 It is clear that the combinatorial structure (described, say, by the graph) of any principal submatrix of such a matrix A is determined by the structure of A. Surprisingly, the same holds also for the combinatorial structure of the Schur complement. Since (compare Remark A.1.26) the Schur complement of the block with indices from the set S is obtained by elimination of unknowns with indices from S from equations with indices from S, this means that the resulting so-called *elimination graph*, i.e. the graph of $G(A(\bar{S}, \bar{S}))$, where $\bar{S}$ is the complement of S in the set of all indices, depends only on $G(A)$ and S, and not on the magnitudes of the entries of A. The description of $G(A(\bar{S}, \bar{S}))$ is in the following theorem:

Theorem A.3.5 *In $G(A(\bar{S}, \bar{S}))$, there is an edge (i, j), $i \in \bar{S}$, $j \in \bar{S}$, $i \neq j$, if and only if there is a path in $G(A)$ from i to j all interior nodes (i.e., different from i and j) of which belong to S.*

Remark A.3.6 Observe the coincidence of several properties with those of positive definite matrices in Theorem A.1.34. In the next theorem, we present an analogy of positive semidefinite matrices.

Theorem A.3.7 *Let A be a matrix in Z of order n. Then the following are equivalent:*

(i) *$A + \varepsilon I$ is an M-matrix for all $\varepsilon > 0$.*
(ii) *All principal minors of A are nonnegative.*
(iii) *The sum of all principal minors of order k is nonnegative for $k = 1, \ldots, n$.*
(iv) *Every real eigenvalue of A is nonnegative.*
(v) *The real part of every eigenvalue of A is nonnegative.*

We denote matrices satisfying these conditions M_0-matrices; they are usually called *possibly singular M-matrices.* Also in this case, the submatrices are M_0-matrices and Schur complements with respect to nonsingular principal submatrices are possibly singular M-matrices. In fact, the following holds:

Theorem A.3.8 *If*

$$A = \begin{bmatrix} A_{11} & A_{12} \\ A_{21} & A_{22} \end{bmatrix}$$

is a singular M-matrix and $Au = 0$, u partitioned as $u = \begin{bmatrix} u_1 \\ u_2 \end{bmatrix}$, then the Schur complement $[A/A_{11}]$ is also a singular M-matrix and $[A/A_{11}]u_2 = 0$.

Theorem A.3.9 *Let A be an irreducible singular M-matrix. Then there exists a positive vector u for which $Au = 0$.*

Remark A.3.10 As in the case of positive definite matrices, an M_0-matrix is an M-matrix if and only if it is nonsingular.

In the next theorem, we list other characteristic properties of the class of real square matrices having just the property (iii) from Theorem A.3.2 or property (ii) from Theorem A.1.34, namely all principal minors are positive. These matrices are called *P-matrices* (cf. [28]).

Theorem A.3.11 *Let A be a real square matrix. Then the following are equivalent:*

(i) *A is a P-matrix, i.e. all principal minors of A are positive.*
(ii) *Whenever D is a nonnegative diagonal matrix of the same order as A, then all principal minors of $A + D$ are different from zero.*
(iii) *For every nonzero vector $x = [x_i]$, there exists an index k such that $x_k(Ax)_k > 0$.*
(iv) *Every real eigenvalue of any principal submatrix of A is positive.*
(v) *The implication*

$$z \geq 0,\ SA^T Sz \leq 0 \ \text{ implies } \ z = 0$$

holds for every diagonal matrix S with diagonal entries 1 or -1.
(vi) *To every diagonal matrix S with diagonal entries 1 or -1, there exists a vector $x \geq 0$ such that $SASx > 0$.*

Let us state some corollaries.

Corollary A.3.12 *Every symmetric P-matrix is positive definite (and, of course, every positive definite matrix is a P-matrix). Every P-matrix in Z is an M-matrix.*

Corollary A.3.13 *If $A \in P$, then there exists a vector $x > 0$ such that $Ax > 0$.*

Corollary A.3.14 *If for a real square matrix A its symmetric part $\frac{1}{2}(A + A^T)$ is positive definite, then $A \in P$.*

An important class closely related to the class of M-matrices is that of *inverse M-matrices*; it consists of real matrices the inverse of which is an M-matrix. By (viii) of Theorem A.3.2 and Corollary A.3.3, the following holds:

Theorem A.3.15 *Let A be an inverse M-matrix. Then A as well as all principal submatrices and all Schur complements of principal submatrices are nonnegative matrices; they even are inverse M-matrices.*

A.4 Hankel matrices

A *Hankel matrix* of order n is a matrix H of the form $H = (h_{i+j})$, $i, j = 0, \dots, n-1$, i.e.

$$H = \begin{bmatrix} h_0 & h_1 & h_2 & \dots & h_{n-1} \\ h_1 & h_2 & \dots & h_{n-1} & h_n \\ h_2 & \dots & & & \\ \dots & & \dots & & \dots \\ h_{n-1} & & \dots & h_{2n-3} & h_{2n-2} \end{bmatrix}.$$

Its entries h_k can be real or complex. Let $\mathcal{H}_n$ denote the class of all $n \times n$ Hankel matrices. Evidently, $\mathcal{H}_n$ is a linear vector space (complex or real) of dimension $2n-1$. It is also clear that an $n \times n$ Hankel matrix has rank one if and only if it is either of the form $\gamma(t^{i+k})$ for γ and t fixed (in general, complex), or if it has a single nonzero entry in the lower-right corner. Hankel matrices play an important role in approximations, investigation of polynomials, etc.

A.5 Projective geometry

For our purpose, we shall introduce, for an integer $n > 1$, the notion of the *real projective n-space P_n*, as the set of points defined as *real homogeneous $(n+1)$-tuples*, i.e. classes of equivalence of all ordered nonzero real $(n+1)$-tuples $(y_1, \dots, y_{n+1})$ under the equivalence

$$(y_1, \ldots, y_{n+1}) \sim (z_1, \ldots, z_{n+1})$$

if and only if the rank of the matrix

$$\begin{bmatrix} y_1 & \ldots & y_{n+1} \\ z_1 & \ldots & z_{n+1} \end{bmatrix} \tag{A.20}$$

is one. This, of course, happens if and only if $z_k = \lambda y_k$ for $k = 1, \ldots, n+1$ for some λ different from zero.

The entries of the $(n+1)$-tuple will be called *(homogeneous) coordinates* of the corresponding point.

Remark A.5.1 *This definition means that P_n is obtained by starting with an $(n+1)$-dimensional real vector space, removing the zero vector and identifying the lines with the points of P_n. Concisely, we should distinguish between the just defined geometric point of P_n and the arithmetic point identified by a chosen $(n+1)$-tuple.*

Usually, we shall denote the points by upper case letters and write simply $Y = (y_1, \ldots, y_{n+1})$ or even $Y = (y_i)$ if $(y_1, \ldots, y_{n+1})$ is some (arithmetic) representative of the point Y. If $Y = (y_i)$ and $Z = (z_i)$ are distinct points, i.e. the matrix (A.20) has rank two, we call the *line determined by Y and Z* and denote by $\mathcal{L}(Y, Z)$ the set of all points some representative of which has the form $(\alpha y_1 + \beta z_1, \ldots, \alpha y_{n+1} + \beta z_{n+1})$ for some real α and β not both equal to zero (observe that then not all entries in this $(n+1)$-tuple are equal to zero).

Furthermore, if $Y = (y_i)$, $Z = (z_i), \ldots, T = (t_i)$ are m points in P_n, we say that they are *linearly independent* (respectively, *linearly dependent*) if the matrix (with m rows)

$$\begin{bmatrix} y_1 & \ldots & y_{n+1} \\ z_1 & \ldots & z_{n+1} \\ \ldots & \ldots & \ldots \\ t_1 & \ldots & t_{n+1} \end{bmatrix}$$

has rank m (respectively, less than m).

It will be convenient to denote by $[y]$, $[z]$ etc. the column vectors with entries (y_i), (z_i) etc. Thus the above condition can also be formulated as the decision of whether (for the matrix transposed to the preceding) the rank $[[y], [z], \ldots, [t]]$ is equal to m or less than m.

The following corollary is self-evident:

Corollary A.5.2 *In P_n, any $n+2$ points are linearly dependent, and there exist $n+1$ linearly independent points.*

Remark A.5.3 Such a set of $n+1$ linearly independent points is called the *basis* of P_n and the number n is the *dimension* of P_n. Of importance for us will also be the sets of $n+2$ points, any $n+1$ of which are linearly independent. We call such a (usually ordered) set a *quasibasis* of P_n.

Let now $Y = (y_i)$, $Z = (z_i)$, ..., $T = (t_i)$ be m linearly independent points in P_n. We denote by $\mathcal{L}[Y, Z, \ldots, T]$ the set of all real *linear combinations* of the points $Y, Z, \ldots, T$, i.e. the set of all points in P_n with the $(n+1)$-tuple of coordinates of the form $(\alpha y_i + \beta z_i + \ldots + \gamma t_i)$ and call it the *linear hull* of the points $Y, Z, \ldots, T$. Since these points are linearly independent, such a linear combination is nonzero if not all coefficients $\alpha, \beta, \ldots, \gamma$ are equal to zero.

A *projective transformation* $\mathcal{T}$ in P_n is a one-to-one (bijective) mapping of P_n into itself which assigns to a point $X = (x_i)$ the point $Y = (y_i)$ defined by

$$[y] = \sigma_X A[x],$$

where A is a (fixed) nonsingular real matrix and σ_X a real number (depending on X). It is clear that projective transformations in P_n form a group with respect to the operation of composition.

We now say that two geometric objects in P_n are *projectively equivalent* if one is obtained as the image of the other in a projective transformation.

Theorem A.5.4 *Any two (ordered) quasibases in P_n are projectively equivalent. In addition, there is a unique projective transformation which maps the first quasibasis into the second.*

We omit the proof but add some comments.

Remark A.5.5 More generally, we can show that the ordered system of m points $Y_1, \ldots, Y_m$, $m \geq n+2$, is projectively equivalent to the ordered system $Z_1, \ldots, Z_m$ if and only if there exist a nonsingular matrix A of order $n+1$ and a nonsingular diagonal matrix D of order m such that for the matrices U, V consisting as above of the arbitrarily chosen column representatives of the points Y_i, Z_i

$$V = AUD \tag{A.21}$$

holds.

Let m be a positive integer less than n. It is obvious that the set of points in P_n with the property that the last $n-m$ coordinates are equal to zero can be identified with the set of all points of a projective space of dimension m having the same first $m+1$ coordinates. This set is the linear hull of the points $B_1 = (1, 0, \ldots, 0)$, ..., $B_{m+1} = (0, \ldots, 0, 1, 0, \ldots, 0)$ (with 1 in the $(m+1)$th place). By the result in Remark A.5.5, we obtain:

Corollary A.5.6 *If $Y = (y_i)$, $Z = (z_i), \ldots, T = (t_i)$ are $m+1$ linearly independent points in $\widehat{P}_n$, then all $(n+1)$-tuples of the form $(\alpha y_1 + \beta z_1 + \cdots +$*

$\gamma t_1, \ldots, \alpha y_{n+1}+\beta z_{n+1}+\cdots+\gamma t_{n+1})$ with $(m+1)$-tuples $(\alpha, \beta, \ldots, \gamma)$ different from zero form an m-dimensional projective space.

Remark A.5.7 This space will be called the *(linear) subspace of P_n determined by the points $Y, Z, \ldots, T$*. However, it is important to realize that the same subspace can be determined by any of its m linearly independent points, i.e. by any of its bases.

If $m = n$, such a subspace is called a *hyperplane* in P_n. It is thus a subspace of maximal dimension in P_n which is different from P_n itself. Similarly, as in vector spaces, such a hyperplane corresponds to a *linear form* which, however, has to be nonzero; in addition, this linear form is determined up to a nonzero multiple.

Let, say, $\langle \alpha, x \rangle$ denote the bilinear form

$$\langle \alpha, x \rangle = \alpha_1 x_1 + \alpha_2 x_2 + \ldots + \alpha_{n+1} x_{n+1}.$$

Here, the symbol x stands for the point $x = (x_1, \ldots, x_{n+1})$, and the symbol α for the hyperplane, which we shall also consider as an $(n+1)$-tuple $(\alpha_1, \ldots, \alpha_{n+1})_d$.

We say that the point x and the hyperplane α are *incident* if $\langle \alpha, x \rangle = 0$. Clearly, this fact does not depend on the choice of nonzero multiples in either variable α or x. Thus α can be considered also as a point of a projective n-dimensional space, say, $P_n^{(d)}$, which we call the *dual* space to P_n. There are two ways to characterize a hyperplane α: either by its equation $\langle \alpha, x \rangle = 0$ describing the set of all points x incident with α, or by the $(n+1)$-tuple of the dual coordinates. The word dual is justified since there is a bilinear form, namely $\langle .,. \rangle$ which satisfies the two properties: to every $\alpha \in P_n^{(d)}$ there exists an element $x_0 \in P_n$ such that $\langle \alpha, x_0 \rangle \neq 0$, and to every $x \in P_n$ there exists an element $\alpha_0 \in P_n^{(d)}$ such that $\langle \alpha_0, x \rangle \neq 0$. Also, P_n is a dual space to $P_n^{(d)}$ since the bilinear form $\langle x, \alpha \rangle_{(d)}$ defined as $\langle \alpha, x \rangle$ again satisfies the two mentioned properties. We are thus allowed to speak about the *equation* of the point, say $Z = (z_i)$, as

$$\xi_1 z_1 + \ldots + \xi_{n+1} z_{n+1} = 0,$$

where the ξs play the role of the coordinates of a variable hyperplane incident with the point Z.

There are simple formulae about how the change of the basis in P_n is reflected by a suitable change of the basis in $P_n^{(d)}$ (cf. [31], p. 30).

Let us only mention that a linear subspace of P_n of dimension m can be determined either by its $m+1$ linearly independent points as their *linear hull* (i.e. the set of all linear combinations of these points), or as the intersection of $n - m - 1$ linearly independent hyperplanes.

We turn now to quadrics in P_n, the next simplest notion. A *quadric* in P_n, more explicitly *quadratic hypersurface* in P_n, is the set $\mathcal{Q}_A$ of all points x in P_n whose coordinates x_i annihilate some quadratic form

$$\sum_{i,k=1}^{n+1} a_{ik} x_i x_k \tag{A.22}$$

not identically equal to zero. As usual, we assume that the coefficients in (A.22) are real and satisfy $a_{ik} = a_{ki}$. The coefficients can thus be written in the form of a real symmetric matrix $A = [a_{ik}]$ and the left-hand side of (A.22) as $[x]^T A[x]$, T meaning as usual the transposition and $[x]$ the column vector with coordinates x_i. The quadric $\mathcal{Q}_A$ is called *nonsingular* if the matrix A is nonsingular, otherwise it is *singular*.

Let $Z = (z_i)$ be a point in P_n. Then two cases can occur. Either all the sums $\sum_k a_{ik} z_k$ are equal to zero, or not. In the first case, we say that Z is a *singular point* of the quadric (which is necessarily singular since $A[z] = 0$ and $[z] \neq 0$). In the second case, the equation

$$\sum_{i,k=1}^{n+1} a_{ik} x_i z_k = 0 \tag{A.23}$$

is an equation of a hyperplane; we call this hyperplane the *polar hyperplane* or simply *polar* of the point Z with respect to the quadric $\mathcal{Q}_A$.

It follows easily from (A.22) and (A.23) that a singular point is always a point of the quadric. If, however, Z is a nonsingular point of the quadric, then the corresponding polar is, since it has the properties required, called the *tangent hyperplane of* $\mathcal{Q}_A$ *at the point* Z. The point Z is then incident with the corresponding tangent hyperplane and in fact this property is (for a nonsingular point of $\mathcal{Q}_A$) characteristic for the polar to be a tangent hyperplane.

We can now formulate the problem to characterize the set of all tangent hyperplanes of $\mathcal{Q}_A$ as a subset of the dual space $P_n^{(d)}$.

Theorem A.5.8 *Let* $\mathcal{Q}_A$ *be a nonsingular quadric with the corresponding matrix* A. *Then the set of all its tangent hyperplanes forms in* $P_n^{(d)}$ *again a nonsingular quadric; this dual quadric corresponds to the matrix* A^{-1}.

Proof. We have to characterize the set of all hyperplanes ξ with the property that

$$[\xi] = A[x]$$

and

$$[x]^T A[x] = 0 \tag{A.24}$$

for some $[x] \neq 0$. Substituting $[x] = A^{-1}[\xi]$ into (A.24) yields

$$[\xi]A^{-1}[\xi] = 0.$$

Since the converse is also true, the proof is complete. □

A singular quadric $\mathcal{Q}_A$ with matrix A of rank r, i.e. $r < n+1$, has *singular* points. The set $\mathcal{S}_Q$ of its singular points is formed by the linear space of those points $X = (x_i)$ whose vector $[x]$ satisfies $A[x] = 0$. Thus the dimension of $\mathcal{S}_Q$ is $n - r$. It is easily seen that if $Y \in \mathcal{Q}_A$ and $Z \in \mathcal{Q}_A$, $Z \neq Y$, then the whole line YZ is contained in $\mathcal{Q}_A$. Thus $\mathcal{Q}_A$ is then a *generalized cone* with the *vertex set* $\mathcal{S}_Q$. Moreover, if $r = 2$, $\mathcal{Q}_A$ is the set of two hyperplanes; its equation is the product of the equations of the hyperplanes and these hyperplanes are distinct. It can, however, happen that the hyperplanes are complex conjugate. If $r = 1$, $\mathcal{Q}_A$ is just one hyperplane; all points of this hyperplane are singular and the equation of $\mathcal{Q}_A$ is the square of the equation of the hyperplane. Whereas the set of all polars of all points with respect to a nonsingular quadric is the set of all hyperplanes, in the case of a singular quadric this set is restricted to those hyperplanes which are incident with the set of singular points $\mathcal{S}_Q$.

Two points $Y = (y_i)$ and $Z = (z_i)$ are called *polar conjugates* with respect to the quadric $\mathcal{Q}_A$ if for the corresponding column vectors $[y]$ and $[z]$, $[y]^T A[z] = 0$. This means that each of the points Y, Z is incident with the polar of the other point, if this polar exists. However, the definition applies also for the case that one or both of the points Y and Z are singular.

Finally, $n + 1$ linearly independent points, any two different of which are polar conjugates with respect of the quadric $\mathcal{Q}_A$, are said to form an *autopolar n-simplex*. In the case that these points are the coordinate points $O_1 = (1, 0, \ldots, 0)$, $O_2 = (0, 1, \ldots, 0)$, ..., $O_{n+1} = (0, 0, \ldots, 1)$, the matrix A has all off-diagonal entries equal to zero. The converse also holds.

All these definitions and facts can be dually formulated for dual quadrics. We must, however, be aware of the fact that only nonsingular quadrics can be considered both as quadrics and dual quadrics at the same time.

Now let a (point) quadric $\mathcal{Q}_A$ with the matrix $A = [a_{ik}]$ and a dual quadric $\mathcal{Q}_\Gamma$ with the matrix $\Gamma = [\gamma_{ik}]$ be given. We say that these quadrics are *apolar* if

$$\sum_{i,k=1}^{n+1} a_{ik}\gamma_{ik} = 0. \tag{A.25}$$

It can be shown that this happens if and only if there exists an autopolar n-simplex of $\mathcal{Q}_A$ with the property that all $n+1$ of its $(n-1)$-dimensional faces are hyperplanes of $\mathcal{Q}_\Gamma$. (Observe that this is true if the simplex is formed by the coordinate points $(1, 0, \ldots, 0)$, etc. since then $a_{ik} = 0$ for all $i, k = 1, \ldots, n+1$, $i \neq k$, as well as $\gamma_{ii} = 0$ for $i = 1, \ldots, n+1$.)

To better understand basic properties of linear and quadratic objects in projective spaces, we shall investigate more thoroughly the case $n = 1$, which is more important than it might seem.

The first fact to be observed is that the hyperplanes as sets of points have also dimension 1. The point $Y = (y_1, y_2)$ is incident only with the dual point $Y^{(d)} = (y_2, -y_1)_d$. The quadrics have ranks 1 or 2. In the first case, such a quadric consists of a single point ("counted twice"), in the second of two distinct points which can also be complex conjugate.

A particularly important notion is that of two harmonic pairs of points. The basis is in the following theorem:

Theorem A.5.9 *Let A, B, C, D be points in P_1. Then the following are equivalent:*

(i) *the quadric of the points A and B and the dual quadric of the points $C^{(d)}$ and $D^{(d)}$ are apolar;*
(ii) *the points C and D are polar conjugates with respect to the quadric formed by the points A and B;*
(iii) *the quadric of the points C and D and the dual quadric of the points $A^{(d)}$ and $B^{(d)}$ are apolar;*
(iv) *the points A and B are polar conjugates with respect to the quadric formed by the points C and D.*

Proof. Let $A = (a_1, a_2)$, etc. Since the quadric of the points A and B has the equation

$$(a_2x_1 - a_1x_2)(b_2x_1 - b_1x_2) = 0$$

and the dual quadric of the points $C^{(d)}$ and $D^{(d)}$ the dual equation

$$(c_1\xi_1 + c_2\xi_2)(d_1\xi_1 + d_2\xi_2) = 0,$$

the condition (i) means that

$$2a_2b_2c_1d_1 - (a_2b_1 + a_1b_2)(c_1d_2 + c_2d_1) + 2a_1b_1c_2d_2 = 0. \tag{A.26}$$

The condition (ii) means that

$$(a_2c_1 - a_1c_2)(b_2d_1 - b_1d_2) + (a_2d_1 - a_1d_2)(b_2c_1 - b_1c_2) = 0. \tag{A.27}$$

Since this condition coincides with (A.26), (i) and (ii) are equivalent. The condition in (iii) is again (A.26), and similarly for (iv). □

If one – and thus all – of the conditions (i)–(iv) is fulfilled, the pairs A, B and C, D are called *harmonic*. Let us add a useful criterion of harmonicity in the case that the points A and B are distinct.

Theorem A.5.10 *Let A, B, and C be points in P_1, A and B distinct. If $C = \alpha A + \beta B$ for some α and β, then $\alpha A - \beta B$ is again a point and this point completes C to a pair harmonic with the pair A and B.*

Proof. Substitute $c_i = \alpha a_i + \beta b_i$, $i = 1, 2$, into (A.27) . We obtain

$$(a_2 b_1 - a_1 b_2)((\alpha a_2 - \beta b_2) d_1 - (\alpha a_1 - \beta b_1) d_2) = 0,$$

which yields the result. □

Remark A.5.11 Some of the points A, B, C, and D may coincide. On the other hand, the pair A, B can be complex conjugate and α and β can also be complex and we still can get a real or complex conjugate pair (only such lead to real quadrics).

The relationship which assigns to every point C in P_1 the point D harmonic to C with respect to some pair A, B is an *involution.* Here again, the pair A, B can be complex conjugate.

Theorem A.5.12 *Such involution is determined by two pairs of points related in this involution (the pairs must not be identical). If these pairs are C, D and E, F, then the relationship between X and Y is obtained from the formula*

$$\det \begin{bmatrix} x_1 y_1 & x_1 y_2 + x_2 y_1 & x_2 y_2 \\ c_1 d_1 & c_1 d_2 + c_2 d_1 & c_2 d_2 \\ e_1 f_1 & e_1 f_2 + e_2 f_1 & e_2 f_2 \end{bmatrix} = 0. \tag{A.28}$$

Proof. This follows from Theorem A.5.9 since (A.28) describes the situation that there is a dual quadric apolar to all three pairs X, Y; C, D; and E, F. Under the stated condition, the last two rows of the determinant are linearly independent. □

For the sake of completeness, let us add the well-known construction of the fourth harmonic point on a line using the plane. If A, B, and C are the given points, we choose a point P not on the line arbitrarily, then Q on PC, different from both P and C. Then construct the intersection points R of PB and QA, and S of PA and QB. The intersection point of RS with the line AB is the fourth harmonic point D.

In Chapter 4, we use the following notion and result:

We call two systems in a projective m-space, with $m+1$ points each, *independent* if for no $k \in \{0, 1, \ldots, m\}$ the following holds: a k-dimensional linear subspace generated by $k+1$ points of any one of the systems contains more that k points of the other system.

Theorem A.5.13 *Suppose that two independent systems with $m+1$ points each in a projective m-space P_m are given. Then there exists at most one nonsingular quadric in P_m for which both systems are autopolar.*

Proof. Choose the points of the first system as vertices $O_1, O_2, \ldots, O_m$ of projective coordinates in P_m, and let $Y_i = (\overset{i}{y}_1, \overset{i}{y}_2, \ldots, \overset{i}{y}_m)$, $i = 1, 2, \ldots, m$ be the points of the second system.

Suppose there are two different nonsingular quadrics having both systems as autopolar. Suppose that

$$a \equiv \sum_{i=1}^{n} a_i\, x_i^2 = 0, \quad b \equiv \sum_{i=1}^{n} b_i\, x_i^2 = 0,$$

are their equations; clearly, a_i, b_i are numbers different from zero, and the rank of the matrix $\begin{bmatrix} a_i \\ b_i \end{bmatrix}$ is 2.

The condition that also the second system is autopolar with respect to both a and b, implies the existence of nonzero numbers $\sigma_1, \sigma_2, \ldots, \sigma_n$, such that for $r = 1, 2, \ldots, n$, we have for all x_is

$$\sum_{i=1}^{n} a_i \overset{r}{y}_i x_i \equiv \sigma_r \sum_{i=1}^{n} b_i \overset{r}{y}_i x_i.$$

Therefore

$$(a_i - \sigma_r b_i)\overset{r}{y}_i = 0 \tag{A.29}$$

for $i, r = 1, \ldots, m$. Define now the following equivalence relation among the indices $1, \ldots, n$: $i \sim j$, if $a_i\, b_j - a_j\, b_i = 0$.

Observe that not all indices $1, \ldots, m$ are in the same class with respect to this equivalence since then the matrix $\begin{bmatrix} a_i \\ b_i \end{bmatrix}$ would have rank less than 2, a contradiction. Denote thus by M_1 the class of all indices equivalent with the index 1, by M_2 the nonvoid set of the remaining indices.

If now $Y_r = (\overset{r}{y}_k)$ is one of the points Y_k, then the nonzero coordinates $\overset{r}{y}_k$ have indices either all from M_1, or all from M_2: Indeed, if for $i \in M_1$, $j \in M_2$, $\overset{r}{y}_i \neq 0$, $\overset{r}{y}_j \neq 0$, then (A.29) would imply

$$a_i = \sigma_r b_i, \qquad a_j = \sigma_r\, b_j,$$

i.e. $i \sim j$, a contradiction.

Denote by p_1, p_2, respectively, the number of points Y_r, having nonzero coordinates in M_1, respectively M_2. We have $p_1 + p_2 = n$; the linear independence of the points Y_r implies that $p_1 \leq s$, $p_2 \leq n - s$, where s is the cardinality of M_1. This means, however, that $p_1 = s$, $p_2 = n - s$. Thus the linear space of dimension $s - 1$, generated by the points O_i for $i \in M_1$, contains s points Y_r, a contradiction with the independence of both systems. □

To conclude this chapter, we investigate the so-called *rational normal curves in P_n*. These are geometric objects whose points are in a one-to-one correspondence with points in a projective line. Because of homogeneity, we shall use forms in two (homogeneous) indeterminates (variables) instead of polynomials in one indeterminate. We can, of course, use similar notions such as factor, divisibility, common divisor, prime forms, etc.

Definition A.5.14 A *rational normal curve* $\mathcal{C}_n$ in P_n is the set of all those points $(x_1, \ldots, x_{n+1})$ which are obtained as image of P_1 in the mapping $f\colon P_1 \to P_n$ given by

$$x_k = f_k(t_1, t_2), \quad k = 1, \ldots, n+1, \tag{A.30}$$

where $f_1(t_1, t_2), \ldots, f_{n+1}(t_1, t_2)$ are linearly independent forms (i.e. homogeneous polynomials) of degree n.

Remark A.5.15 For $n = 1$, we obtain the whole line P_1. As we shall see, for $n = 2$, $\mathcal{C}_2$ is a nonsingular conic. In general, it is a curve of degree n (in the sense that it has n points in common with every hyperplane of P_n if appropriate multiplicities of the common points are defined). Of course, (A.30) are the parametric equations of $\mathcal{C}_n$.

Theorem A.5.16 *$\mathcal{C}_n$ has the following properties:*

(i) *it contains $n+1$ linearly independent points (which means that it is not contained in any hyperplane);*
(ii) *in an appropriate basis of P_n, its parametric equations are*

$$x_k = t_1^{n+1-k} t_2^{k-1}, \quad k = 1, \ldots, n+1; \tag{A.31}$$

(iii) *for $n \geq 2$, $\mathcal{C}_n$ is the intersection of $n-1$ linearly independent quadrics.*

Proof. The property (i) just rewords the condition that the forms f_k are linearly independent. Also, if we express these forms explicitly

$$f_k(t_1, t_2) = f_{k,0} t_1^n + f_{k,1} t_1^{n-1} t_2 + \ldots + f_{k,n} t_2^n, \quad k = 1, 2, \ldots, n+1,$$

then the matrix Φ of the coefficients

$$\Phi = (f_{k,l}), \quad k = 1, \ldots, n+1, \; l = 0, \ldots, n$$

is nonsingular. This implies (ii) since the transformation of the coordinates with the matrix Φ^{-1} will bring the coefficient matrix to the identity matrix as in (A.31).

To prove (iii), it suffices to choose as $\mathcal{C}_n$ in the form (A.31) the quadrics with equations

$$x_1 x_3 - x_2^2 = 0, \; x_2 x_4 - x_3^2 = 0, \; \ldots, \; x_{n-1} x_{n+1} - x_n^2 = 0. \tag{A.32}$$

Clearly every point of $\mathcal{C}_n$ is contained in all the quadrics in (A.32). Conversely, let a point $Y = (y_1, \ldots, y_{n+1})$ be contained in all these quadrics. If $y_1 = 0$, Y is the point $(0, 0, \ldots, 0, 1)$ which belongs to $\mathcal{C}_n$ for $t_1 = 0$, $t_2 = 1$. If $y_1 \neq 0$, set $t = y_2/y_1$. Then $y_2 = ty_1$, $y_3 = t^2y_1$, …, $y_{n+1} = t^n y_1$, which means that Y corresponds to $t_1 = 1$, $t_2 = t$. □

Corollary A.5.17 *C_2 is a nonsingular conic.*

Proof. Indeed, it is – in the form (A.32) – the conic with equation $x_1x_3 - x_2^2 = 0$ and this conic is nonsingular. □

Theorem A.5.18 *Any $n+1$ distinct points of C_n are linearly independent.*

Proof. We can bring C_n to the form (A.31) by choosing an appropriate coordinate system. Since the given points have distinct ratios of parameters, the matrix of coordinates being essentially the Vandermonde matrix with nonproportional columns is nonsingular (cf. [28]). □

Theorem A.5.19 *Any two rational normal curves, each in a real projective space of dimension n, are projectively equivalent. Any two points of such a curve are projectively equivalent as well.*

Proof. The first assertion follows from the fact that every such curve is projectively equivalent to the curve of the form (A.31). The second is obtained by a suitable linear transformation of the homogeneous parameters. □

References

[1] L. M. Blumenthal: *Theory and Applications of Distance Geometry.* Oxford, Clarendon Press, 1953.

[2] E. Egerváry: On orthocentric simplexes. *Acta Math. Szeged IX* (1940), 218–226.

[3] M. Fiedler: Geometrie simplexu I. *Časopis pěst. mat.* 79 (1954), 270–297.

[4] M. Fiedler: Geometrie simplexu II. *Časopis pěst. mat.* 80 (1955), 462–476.

[5] M. Fiedler: Geometrie simplexu III. *Časopis pěst. mat.* 81 (1956), 182–223.

[6] M. Fiedler: Über qualitative Winkeleigenschaften der Simplexe. *Czechosl. Math. J.* 7(82) (1957), 463–478.

[7] M. Fiedler: Einige Sätze aus der metrischen Geometrie der Simplexe in Euklidischen Räumen. In: *Schriftenreihe d. Inst. f. Math. DAW*, Heft 1, Berlin (1957), 157.

[8] M. Fiedler: A note on positive definite matrices. (Czech, English summary.) *Czechosl. Math. J.* 10(85) (1960), 75–77.

[9] M. Fiedler: Über eine Ungleichung für positive definite Matrizen. *Mathematische Nachrichten* 23 (1961), 197–199.

[10] M. Fiedler: Über die qualitative Lage des Mittelpunktes der umgeschriebenen Hyperkugel im n-Simplex. *Comm. Math. Univ. Carol.* 2(1) (1961), 3–51.

[11] M. Fiedler: Über zyklische n-Simplexe und konjugierte Raumvielecke. *Comm. Math. Univ. Carol.* 2(2) (1961), 3–26.

[12] M. Fiedler, V. Pták: On matrices with non-positive off-diagonal elements and positive principal minors. *Czechosl. Math. J.* 12(87) (1962), 382–400.

[13] M. Fiedler: Hankel matrices and 2-apolarity. *Notices AMS* 11 (1964), 367–368.

[14] M. Fiedler: Relations between the diagonal elements of two mutually inverse positive definite matrices. *Czechosl. Math. J.* 14(89) (1964), 39–51.

[15] M. Fiedler: Some applications of the theory of graphs in the matrix theory and geometry. In: *Theory of Graphs and Its Applications. Proc. Symp. Smolenice 1963*, Academia, Praha (1964), 37–41.

[16] M. Fiedler: Matrix inequalities. *Numer. Math.* 9 (1966), 109–119.

[17] M. Fiedler: Algebraic connectivity of graphs. *Czechosl. Math. J.* 23(98) (1973), 298–305.

[18] M. Fiedler: Eigenvectors of acyclic matrices. *Czechosl. Math. J.* 25(100) (1975), 607–618.

[19] M. Fiedler: A property of eigenvectors of nonnegative symmetric matrices and its application to graph theory. *Czechosl. Math. J.* 25(100) (1975), 619–633.

[20] M. Fiedler: Aggregation in graphs. In: *Coll. Math. Soc. J. Bolyai, 18. Combinatorics.* Keszthely (1976), 315–330.

[21] M. Fiedler: Laplacian of graphs and algebraic connectivity. In: *Combinatorics and Graph Theory, Banach Center Publ. vol. 25*, PWN, Warszava (1989), 57–70.

[22] M. Fiedler: A geometric approach to the Laplacian matrix of a graph. In: *Combinatorial and Graph-Theoretical Problems in Linear Algebra* (R. A. Brualdi, S. Friedland, V. Klee, editors), Springer, New York (1993), 73–98.

[23] M. Fiedler: Structure ranks of matrices. *Linear Algebra Appl.* 179 (1993), 119–128.

[24] M. Fiedler: Elliptic matrices with zero diagonal. *Linear Algebra Appl.* 197, 198 (1994), 337–347.

[25] M. Fiedler: Moore–Penrose involutions in the classes of Laplacians and simplices. *Linear Multilin. Algebra* 39 (1995), 171–178.

[26] M. Fiedler: Some characterizations of symmetric inverse M-matrices. *Linear Algebra Appl.* 275–276 (1998), 179–187.

[27] M. Fiedler: Moore-Penrose biorthogonal systems in Euclidean spaces. *Linear Algebra Appl.* 362 (2003), 137–143.

[28] M. Fiedler: *Special Matrices and Their Applications in Numerical Mathematics*, 2nd edn, Dover Publ., Mineola, NY (2008).

[29] M. Fiedler, T. L. Markham: Rank-preserving diagonal completions of a matrix. *Linear Algebra Appl.* 85 (1987), 49–56.

[30] M. Fiedler, T. L. Markham: A characterization of the Moore–Penrose inverse. *Linear Algebra Appl.* 179 (1993), 129–134.

[31] R. A. Horn, C. A. Johnson: *Matrix Analysis*, Cambridge University Press, New York, NY (1985).

[32] D. J. H. Moore: A geometric theory for electrical networks. Ph.D. Thesis, Monash. Univ., Australia (1968).

Index